ASE Test Preparation Series

Automobile Test

Engine Performance (Test A8)

4th Edition

THOMSON
DELMAR LEARNING™

Australia Canada Mexico Singapore Spain United Kingdom United States

Thomson Delmar Learning's ASE Test Preparation Series
Automobile Test for Engine Performance (Test A8), 4th Edition

Vice President, Technology Professional Business Unit:
Gregory L. Clayton

Product Development Manager:
Kristen Davis

Product Manager:
Kim Blakey

Editorial Assistant:
Vanessa Carlson

Director of Marketing:
Beth A. Lutz

Marketing Specialist:
Brian McGrath

Marketing Coordinator:
Marissa Maiella

Production Manager:
Andrew Crouth

Production Editor:
Kara A. DiCaterino

Senior Project Editor:
Christopher Chien

XML Architect:
Jean Kaplansky

Cover Design:
Michael Egan

Cover Images:
Portion courtesy of DaimlerChrysler Corporation

Printed in the United States of America.
1 2 3 4 5 XXX 10 09 08 07 06

For more information contact:
Thomson Delmar Learning
Executive Woods
5 Maxwell Drive, PO Box 8007,
Clifton Park, NY 12065-8007
Or find us on the
World Wide Web
at www.delmarlearning.com
or www.trainingbay.com

ISBN: 1-4180-3885-7

NOTICE TO THE READER

Contents

Section 5 Sample Test for Practice

Section 6 Additional Test Questions for Practice

Section 7 Appendices

Preface

Delmar Learning is very pleased that you have chosen our ASE Test Preparation Series to prepare yourself for the automotive ASE Examination. These guides are available for all of the automotive areas including A1–A8, the L1 Advanced Diagnostic Certification, the P2 Parts Specialist, the C1 Service Consultant and the X1 Undercar Specialist. These guides are designed to introduce you to the Task List for the test you are preparing to take, give you an understanding of what you are expected to be able to do in each task, and take you through sample test questions formatted in the same way the ASE tests are structured.

If you have a basic working knowledge of the discipline you are testing for, you will find Delmar Learning's ASE Test Preparation Series to be an excellent way to understand the "must know" items to pass the test. These books are not textbooks. Their objective is to prepare the technician who has the requisite experience and schooling to challenge ASE testing. It cannot replace the hands-on experience or the theoretical knowledge required by ASE to master vehicle repair technology. If you are unable to understand more than a few of the questions and their explanations in this book, it could be that you require either more shop-floor experience or further study. Some resources that can assist you with further study are listed on the rear cover of this book.

Each book begins with an item-by-item overview of the ASE Task List with explanations of the minimum knowledge you must possess to answer questions related to the task. Following that there are 2 sets of sample questions followed by an answer key to each test and an explanation of the answers to each question. A few of the questions are not strictly ASE format but were included because they help teach a critical concept that will appear on the test. We suggest that you read the complete Task List Overview before taking the first sample test. After taking the first test, score yourself and read the explanation to any questions that you were not sure about, including the questions you answered correctly. Each test question has a reference back to the related task or tasks that it covers. This will help you to go back and read over any area of the task list that you are having trouble with. Once you are satisfied that you have all of your questions answered from the first sample test, take the additional tests and check them. If you pass these tests, you will be prepared to do well on the ASE test.

Our Commitment to Excellence

The 4th edition of Delmar Learning's ASE Test Preparation Series has been through a major revision with extensive updates to the ASE's task lists, test questions, and answers and explanations. Delmar Learning has sought out the best technicians in the country to help with the updating and revision of each of the books in the series.

About the Series Advisor

To promote consistency throughout the series, a series advisor took on the task of reading, editing, and helping each of our experts give each book the highest level of accuracy possible. Dan Perrin has served in the role of Series Advisor for the 4th edition of the ASE Test Preparation Series. Dan began ASE testing with the first series of tests in 1972 and has been continually certified ever since. He holds ASE master status in automotive, truck, collision, and machinist. He is also L1, L2, and alternated fuels certified, along with some others that have expired. He has been an automotive educator since 1979, having taught at the secondary, post-secondary, and industry levels. His service includes participation on boards that include the North American Council of Automotive Teachers (NACAT), the Automotive Industry Planning Council (AIPC), and the National Automotive Technicians Education Foundation (NATEF). Dan currently serves as the Executive Manager of NACAT and Director of the NACAT Education Foundation.

Thanks for choosing Delmar Learning's ASE Test Preparation Series. All of the writers, editors, Delmar Staff, and myself have worked very hard to make this series second to none. I know you are going to find this book accurate and easy to work with. It is our objective to constantly improve our product at Delmar by responding to feedback.

If you have any questions concerning the books in this series, you can email me at: autoexpert@trainingbay.com.

Dan Perrin
Series Advisor

1 The History and Purpose of ASE

ASE began as the National Institute for Automotive Service Excellence (NIASE). It was founded as a non-profit independent entity in 1972 by a group of industry leaders with the single goal of providing a means for consumers to distinguish between incompetent and competent technicians. It accomplishes this goal by testing and certification of repair and service professionals. From this beginning it has evolved to be known simply as ASE (Automotive Service Excellence) and today offers more than 40 certification exams in automotive, medium/heavy duty truck, collision, engine machinist, school bus, parts specialist, automobile service consultant, and other industry-related areas. At this time there are more than 400,000 professionals with current ASE certifications. These professionals are employed by new car and truck dealerships, independent garages, fleets, service stations, franchised service facilities, and more. ASE continues its mission by also providing information that helps consumers identify repair facilities that employ certified professionals through its Blue Seal of Excellence Recognition Program. Shops that have a minimum of 75% of their repair technicians ASE certified and meet other criteria can apply for and receive the Blue Seal of Excellence Recognition from ASE.

ASE recognized that educational programs serving the service and repair industry also needed a way to be recognized as having the faculty, facilities, and equipment to provide a quality education to students wanting to become service professionals. Through the combined efforts of ASE, industry, and education leaders, the non-profit National Automotive Technicians Education Foundation (NATEF) was created to evaluate and recognize training programs. Today more than 2000 programs are ASE certified under the standards set by the service industry. ASE/NATEF also has a certification of industry (factory) training program known as CASE. CASE stands for Continuing Automotive Service Education and recognizes training provided by replacement parts manufacturers as well as vehicle manufacturers.

ASE certification testing is administered by the American College Testing (ACT). Strict standards of security and supervision at the test centers insure that the technician who holds the certification earned it. Additionally ASE certification also requires that the person passing the test to be able to demonstrate that they have two years of work experience in the field before they can be certified. Test questions are developed by industry experts that are actually working in the field being tested. There is more detail on how the test is developed and administered in the next section. Paper and pencil tests are administered twice a year at over seven hundred locations in the United States. Computer based testing is now also available with the benefit of instant test results at certain established test centers. The certification is valid for five years and can be recertified by retesting. So that consumers can recognize certified technicians, ASE issues a jacket patch, certificate, and wallet card to certified technicians and makes signs available to facilities that employ ASE certified technicians.

You can contact ASE at any of the following:

National Institute for Automotive Service Excellence
101 Blue Seal Drive S.E.
Suite 101
Leesburg, VA 20175
Telephone 703-669-6600
FAX 703-669-6123
www.ase.com

2 Take and Pass Every ASE Test

Participating in an Automotive Service Excellence (ASE) voluntary certification program gives you a chance to show your customers that you have the "know-how" needed to work on today's modern vehicles. The ASE certification tests allow you to compare your skills and knowledge to the automotive service industry's standards for each specialty area.

If you are the "average" automotive technician taking this test, you are in your mid-thirties and have not attended school for about fifteen years. That means you probably have not taken a test in many years. Some of you, on the other hand, have attended college or taken postsecondary education courses and may be more familiar with taking tests and with test-taking strategies. There is, however, a difference in the ASE test you are preparing to take and the educational tests you may be accustomed to.

How are the tests administered?

ASE test are administered at over 750 test sites in local communities. Paper and pencil tests are the type most widely available to technicians. Each tester is given a booklet containing questions with charts and diagrams where required. You can mark in this test booklet but no information entered in the booklet is scored. Answers are recorded on a separate answer sheet. You will enter your answers, using a number 2 pencil only. ASE recommends you bring four sharpened number 2 pencils that have erasers. Answer choices are recorded by coloring in the blocks on the answer sheet. The answer sheets are scanned electronically and the answers tabulated. For test security, test booklets include randomly generated questions. Your answer key must be matched to the proper booklet so it is important to correctly enter the booklet serial number on the answer sheet. All instructions are printed on the test materials and should be followed carefully.

ASE has introduced Computer Based Testing (CBT) at some locations. While the test content is the same for both testing methods the CBT tests have some unique requirements and advantages. It is strongly recommended that technicians considering the CBT tests go the ASE web page at www.ASE.com and review the conditions and requirements for this type of test. There is a demonstration of a CBT that allows you to experience this type of test before you register. Some technicians find this style of testing provides an advantage, while others find operating the computer a distraction. One significant benefit of CBT is the availability of instant results. You can receive your test results before you leave the test center. CBT testing also offers increased flexibility in scheduling. The cost for taking CBTs is slightly higher than paper and pencil tests and the number of testing sites is limited. The first time test taker may be more comfortable with the paper and pencil tests but technicians now have a choice.

Who Writes the Questions?

The questions are written by service industry experts in the area being tested. Each area will have its own technical experts. Questions are entirely job related. They are designed to test the skills you need to be a successful technician. Theoretical knowledge is important and necessary to answer the questions, but the ability to apply that knowledge is the basis of ASE test questions.

Each question has its roots in an ASE "item-writing" workshop where service representatives from automobile manufacturers (domestic and import), aftermarket parts and equipment manufacturers, working technicians, and vocational educators meet in a workshop setting to share ideas and translate them into test questions. Each test question written by these experts must survive review by all members of the group.

The questions are written to deal with practical application of soft skills and system knowledge experienced by technicians in their day-to-day work.

All questions are pre-tested and quality-checked on a national sample of technicians. Those questions that meet ASE standards of quality and accuracy are included in the scored sections of the tests; the "rejects" are sent back to the drawing board or discarded altogether.

Each certification test is made up of between forty and eighty multiple-choice questions.

Note: Each test could contain additional questions that are included for statistical research purposes only. Your answers to these questions will not affect your score, but since you do not know which ones they are, you should answer all questions on the test. The five-year Recertification Test will cover the same content areas as those listed above. However, the number of questions in each content area of the Recertification Test will be reduced by about one-half.

Objective Tests

A test is called an objective test if the same standards and conditions apply to everyone taking the test and there is only one correct answer to each question.

Objective tests primarily measure your ability to recall information. A well-designed objective test can also test your ability to understand, analyze, interpret, and apply your knowledge. Objective tests include true-false, multiple choice, fill in the blank, and matching questions. ASE's tests consist exclusively of four-part multiple-choice objective questions.

The following are some strategies that may be applied to your tests.

Before beginning to take an objective test, quickly look over the test to determine the number of questions, but do not try to read through all of the questions. In an ASE test, there are usually between forty and eighty questions, depending on the subject. Read through each question before marking your answer. Answer the questions in the order they appear on the test. Leave the questions blank that you are not sure of and move on to the next question. You can return to those unanswered questions after you have finished the others. They may be easier to answer at a later time after your mind has had additional time to consider them on a subconscious level. In addition, you might find information in other questions that will help you recall the answers to some of them.

Do not be obsessed by the apparent pattern of responses. For example, do not be influenced by a pattern like **D, C, B, A, D, C, B, A** on an ASE test.

There is also a lot of folk wisdom about taking objective tests. For example, there are those who would advise you to avoid response options that use certain words such as *all, none, always, never, must,* and *only,* to name a few. This, they claim, is because nothing in life is exclusive. They would advise you to choose response options that use words that allow for some exception, such as *sometimes, frequently, rarely, often, usually, seldom,* and *normally.* They would also advise you to avoid the first and last option (A and D) because test writers, they feel, are more comfortable if they put the correct answer in the middle (B and C) of the choices. Another recommendation often offered is to select the option that is either shorter or longer than the other three choices because it is more likely to be correct. Some would advise you to never change an answer since your first intuition is usually correct.

Although there may be a grain of truth in this folk wisdom, ASE test writers try to avoid them and so should you. There are just as many **A** answers as there are **B** answers, just as many **D** answers as **C** answers. As a matter of fact, ASE tries to balance the answers at about 25 percent per choice **A, B, C,** and **D.** There is no intention to use "tricky" words, such as outlined above. Put no credence in the opposing words "sometimes" and "never," for example.

Multiple-choice tests are sometimes challenging because there are often several choices that may seem possible, and it may be difficult to decide on the correct choice. The best strategy, in this case, is to first determine the correct answer before looking at the options. If you see the answer you decided on, you should still examine the options to make sure that none seem more correct than yours. If you do not know or are not sure of the answer, read each option very carefully and try to eliminate those options that you know to be wrong. That way, you can often arrive at the correct choice through a process of elimination.

If you have gone through all of the test and you still do not know the answer to some of the questions, then guess. Yes, guess. You then have at least a 25 percent chance of being correct. If you leave the question blank, you have no chance. Your score is based on the number of questions answered correctly.

Preparing for the Exam

The main reason we have included so many sample and practice questions in this guide is, simply, to help you learn what you know and what you don't know. We recommend that you work your way through each question in this book. Before doing this, carefully look through Section 3; it contains a description and explanation of the question types you'll find on an ASE exam.

Once you understand what the questions will look like, move to the sample test. Answer one of the sample questions (Section 5) then read the explanation (Section 7) to the answer for that question. If you don't feel you understand the reasoning for the correct answer, go back and read the overview (Section 4) for the task that is related to that question. If you still don't feel you have a solid understanding of the material, identify a good source of information on the topic, such as a textbook, and do some more studying.

After you have completed all of the sample test items and reviewed your answers, move to the additional questions (Section 6). This time answer the questions as if you were taking an actual test. Do not use any reference or allow any interruptions in order to get a feel for how you will do on an actual test. Once you have answered all of the questions, grade your results using the answer key in Section 7. For every question that you gave a wrong answer to, study the explanations to the answers and/or the overview of the related task areas. Try to determine the root cause for your missing the question. The easiest thing to correct is learning the correct technical content. The hardest thing to correct is behaviors that lead you to a wrong conclusion. If you knew the information but still got it wrong there is a behavior problem that will need to be corrected. An example would be reading too quickly and skipping over words that affect your reasoning. If you can identify what you did that caused you to answer the question incorrectly you can eliminate that cause and improve your score. Here are some basic guidelines to follow while preparing for the exam:

- Focus your studies on those areas you are weak in.
- Be honest with yourself while determining if you understand something.
- Study often but in short periods of time.
- Remove yourself from all distractions while studying.
- Keep in mind the goal of studying is not just to pass the exam, the real goal is to learn!
- Prepare physically by getting a good night's rest before the test and eat meals that provide energy but do not cause discomfort.
- Arrive early to the test site to avoid long waits as test candidates check in and to allow all of the time available for your tests.

During the Test

On paper and pencil tests you will be placing your answers on a sheet where you will be required to color in your answer choice. Stray marks or incomplete erasures may be picked up as an answer by the electronic reader, so be sure only your answers end up on the sheet. One of the biggest problems an adult faces in test taking, it seems, is placing the answer in the correct spot on the answer sheet. Make certain that you mark your answer for, say, question 21, in the space on the answer sheet designated for the answer for question 21. A correct response in the wrong line will probably result in two questions being marked wrong, one with two answers (which could include a correct answer but will be scored wrong) and the other with no answer. Remember, the answer sheet on the written test is machine scored and can only "read" what you have colored in.

If you finish answering all of the questions on a test and have remaining time, go back and review the answers to those questions that you were not sure of. You can often catch careless errors by using the remaining time to review your answers. Carefully check your answer sheet for blank answer blocks or missing information.

At practically every test, some technicians will invariably finish ahead of time and turn their papers in long before the final call. Some technicians may be doing recertification tests and others may be taking fewer tests than you. Do not let them distract or intimidate you.

It is not wise to use less than the total amount of time that you are allotted for a test. If there are any doubts, take the time for review. Any product can usually be made better with some additional effort. A test is no exception. It is not necessary to turn in your test paper until you are told to do so.

Testing Time Length

An ASE written test session is four hours. You may attempt from one to a maximum of four tests in one session. It is recommended, however, that no more than a total of 225 questions be attempted at any test session. This will allow for just over one minute for each question.

Visitors are not permitted at any time. If you wish to leave the test room, for any reason, you must first ask permission. If you finish your test early and wish to leave, you are permitted to do so only during specified dismissal periods.

You should monitor your progress and set an arbitrary limit to how much time you will need for each question. This should be based on the number of questions you are attempting. It is suggested that you wear a watch because some facilities may not have a clock visible to all areas of the room.

Computer-Based Tests are allotted a testing time according to the number of questions ranging from one half hour to one and one half hours. Advanced level tests are allowed two hours. This time is by appointment and you should be sure to be on time to insure that you have all of the time allocated. If you arrive late for a CBT test appointment you will only have the amount of time remaining on your appointment.

Your Test Results!

You can gain a better perspective about tests if you know and understand how they are scored. ASE's tests are scored by American College Testing (ACT), a nonpartial, unbiased organization having no vested interest in ASE or in the automotive industry.

Each question carries the same weight as any other question. For example, if there are fifty questions, each is worth 2 percent of the total score. The passing grade is 70 percent. That means you must correctly answer thirty-five of the fifty questions to pass the test.

The test results can tell you:

- where your knowledge equals or exceeds that needed for competent performance, or
- where you might need more preparation.

Your ASE test score report is divided into content areas and will show the number of questions in each content area and how many of your answers were correct. These numbers provide information about your performance in each area of the test. However, because there may be a different number of questions in each content area of the test, a high percentage of correct answers in an area with few questions may not offset a low percentage in an area with many questions.

It should be noted that one does not "fail" an ASE test. The technician who does not pass is simply told "More Preparation Needed." Though large differences in percentages may indicate problem areas, it is important to consider how many questions were asked in each area. Since each test evaluates all phases of the work involved in a service specialty, you should be prepared in each area. A low score in one area could keep you from passing an entire test.

There is no such thing as average. You cannot determine your overall test score by adding the percentages given for each task area and dividing by the number of areas. It doesn't work that way because there generally are not the same number of questions in each task area. A task area with twenty questions, for example, counts more toward your total score than a task area with ten questions.

Your test report should give you a good picture of your results and a better understanding of your strengths and weaknesses for each task area.

If you fail to pass the test, you may take it again at any time it is scheduled to be administered. You are the only one who will receive your test score. Test scores will not be given over the telephone by ASE nor will they be released to anyone without your written permission.

3 Types of Questions on an ASE Exam

ASE certification tests are often thought of as being tricky. They may seem to be tricky if you do not completely understand what is being asked. The following examples will help you recognize certain types of ASE questions and avoid common errors.

Paper-and-pencil tests and computer-based test questions are identical in content and difficulty. Most initial certification tests are made up of forty to eighty multiple-choice questions. Multiple-choice questions are an efficient way to test knowledge. To answer them correctly, you must think about each choice as a possibility, and then choose the one that best answers the question. To do this, read each word of the question carefully. Do not assume you know what the question is about until you have finished reading it.

About 10 percent of the questions on an actual ASE exam will use an illustration. These drawings contain the information needed to correctly answer the question. The illustration must be studied carefully before attempting to answer the question. Often, technicians look at the possible answers then try to match up the answers with the drawing. Always do the opposite; match the drawing to the answers. When the illustration is showing an electrical schematic or another system in detail, look over the system and try to figure out how the system works before you look at the question and the possible answers.

Multiple-Choice Questions

The most common type of question used on ASE Tests is the multiple-choice question. This type of question contains three "distracters" (wrong answers) and one "key" (correct answer). When the questions are written effort is made to make the distracters plausible to draw an inexperienced technician to one of them. This type of question gives a clear indication of the technician's knowledge. Using multiple criteria including cross-sections by age, race, and other background information, ASE is able to guarantee that a question does not bias for or against any particular group. A question that shows bias toward any particular group is discarded. If you encounter a question that you are unsure of, reverse engineer it by eliminating the items that it cannot be. For example:

A rocker panel is a structural member of which vehicle construction type?

A. Front-wheel drive
B. Pickup truck
C. Unibody
D. Full-frame

Analysis:

This question asks for a specific answer. By carefully reading the question, you will find that it asks for a construction type that uses the rocker panel as a structural part of the vehicle.

Answer A is wrong. Front-wheel drive is not a vehicle construction type.
Answer B is wrong. A pickup truck is not a type of vehicle construction.
Answer C is correct. Unibody design creates structural integrity

by welding parts together, such as the rocker panels, but does not require exterior cosmetic panels installed for full strength.
Answer D is wrong. Full-frame describes a body-over-frame construction type that relies on the frame assembly for structural integrity.

Therefore, the correct answer is C. If the question was read quickly and the words "construction type" were passed over, answer A may have been selected.

EXCEPT Questions

Another type of question used on ASE tests has answers that are all correct except one. The correct answer for this type of question is the answer that is wrong. The word "**EXCEPT**" will always be in capital letters. You must identify which of the choices is the wrong answer. If you read quickly through the question, you may overlook what the question is asking and answer the question with the first correct statement. This will make your answer wrong. An example of this type of question and the analysis is as follows:

All of the following are tools for the analysis of structural damage **EXCEPT:**

A. height gauge
B. tape measure.
C. dial indicator.
D. tram gauge.

Analysis:

The question really requires you to identify the tool that is not used for analyzing structural damage. All tools given in the choices are used for analyzing structural damage except one. This question presents two basic problems for the test-taker who reads through the question too quickly. It may be possible to read over the word "**EXCEPT**" in the question or not think about which type of damage analysis would use answer C. In either case, the correct answer may not be selected. To correctly answer this question, you should know what tools are used for the analysis of structural damage. If you cannot immediately recognize the incorrect tool, you should be able to identify it by analyzing the other choices.

Answer A is wrong. A height gauge may be used to analyze structural damage.
Answer B is wrong. A tape measure may be used to analyze structural damage.
Answer C is correct. A dial indicator may be used as a damage analysis tool for moving parts, such as wheels, wheel hubs, and axle shafts, but would not be used to measure structural damage.
Answer D is wrong. A tram gauge is used to measure structural damage.

Technician A, Technician B Questions

The type of question that is most popularly associated with an ASE test is the "Technician A says . . . Technician B says . . . Who is right?" type. In this type of question, you must identify the correct statement or statements. To answer this type of question correctly, you must carefully read each technician's statement and judge it on its own merit to determine if the statement is true.

Sometimes this type of question begins with a statement about some analysis or repair procedure. This is often referred to as the stem of the question and provides the setup or background information required to understand the conditions the question is based on. This is followed by two statements about the cause of the concern, proper inspection, identification, or repair choices. You are asked whether the first statement, the second statement, both statements, or neither statement is correct. Analyzing this type of question is a little easier than the other types because there are only two ideas to consider although there are still four choices for an answer.

Technician A, Technician B questions are really double true or false questions. The best way to analyze this kind of question is to consider each technician's statement separately. Ask yourself, is A true or false? Is B true or false? Then select your answer from the four choices. An important point to remember is that an ASE Technician A, Technician B question will never have Technician A and B directly disagreeing with each other. That is why you must evaluate each statement independently.

An example of this type of question and the analysis of it follows.

A vehicle comes into the shop with a gas gauge that will not register above one half full. When the sending unit circuit is disconnected the gauge reads empty and when it is connected to ground the gauge goes to full. Technician A says that the sending unit is shorted to ground. Technician B says the gauge circuit is working and the sending unit is likely the problem. Who is right?

A. A only
B. B only
C. Both A and B
D. Neither A nor B

Analysis:

Reading of the stem of the question sets the conditions of the customer concern and establishes what information is gained from testing. General knowledge of gauge circuits and test procedures are needed to correctly evaluate the technician's conclusions. Note: Avoid being distracted by experience with unusual or problem vehicles that you may have worked on, Other technicians taking the same test do not have that knowledge, so it should not be used as the basis of your answers.

Technician A is wrong because a shorted to ground sending unit would produce a gauge reading equivalent to the test conditions of a grounding the circuit and produce a full reading.
Technician B is correct because the gauge spans when going from an open circuit to a completely
grounded circuit. This would tend to indicate that the problem had to be in the sending unit.
Answer C is not correct. Both technicians are identifying the problem as a sending unit but technician A qualified the problem as a specific type of failure (grounded) that would not have caused the symptoms of the vehicle.
Answer D is not correct because technician B's diagnosis is a possible cause of the conditions identified.

Most-Likely Questions

Most-Likely questions are somewhat difficult because only one choice is correct while the other three choices are nearly correct. An example of a Most-Likely-cause question is as follows:

The Most-Likely cause of reduced turbocharger boost pressure may be a:

A. wastegate valve stuck closed.
B. wastegate valve stuck open.
C. leaking wastegate diaphragm.
D. disconnected wastegate linkage.

Analysis:

Answer A is wrong. A wastegate valve stuck closed increases turbocharger boost pressure.
Answer B is correct. A wastegate valve stuck open decreases turbocharger boost pressure.
Answer C is wrong. A leaking wastegate valve diaphragm increases turbocharger boost pressure.

Answer D is wrong. A disconnected wastegate valve linkage will increase turbocharger boost pressure.

LEAST-Likely Questions

Notice that in Most-Likely questions there is no capitalization. This is not so with LEAST-Likely type questions. For this type of question, look for the choice that would be the LEAST-Likely cause of the described situation. Read the entire question carefully before choosing your answer. An example is as follows:

What is the LEAST-Likely cause of a bent pushrod?

A. Excessive engine speed
B. A sticking valve
C. Excessive valve guide clearance
D. A worn rocker arm stud

Analysis:

Answer A is wrong. Excessive engine speed may cause a bent pushrod.
Answer B is wrong. A sticking valve may cause a bent pushrod.
Answer C is correct. Excessive valve clearance will not generally cause a bent pushrod.
Answer D is wrong. A worn rocker arm stud may cause a bent pushrod.

You should avoid relating questions to those unusual situations that you may have encountered and answer based on the technical and mechanical possibilities.

Summary

There are no four-part multiple-choice ASE questions having "none of the above" or "all of the above" choices. ASE does not use other types of questions, such as fill-in-the-blank, completion, true-false, word-matching, or essay. ASE does not require you to draw diagrams or sketches. If a formula or chart is required to answer a question, it is provided for you. There are no ASE questions that require you to use a pocket calculator.

4 Overview of the Task List

Engine Performance (Test A8)

The following section includes the task areas and task lists for this test and a written overview of the topics covered in the test.

The task list describes the actual work you should be able to do as a technician that you will be tested on by the ASE. This is your key to the test and you should review this section carefully. We have based our sample test and additional questions upon these tasks, and the overview section will also support your understanding of the task list. ASE advises that the questions on the test may not equal the number of tasks listed; the task lists tell you what ASE expects you to know how to do and be ready to be tested upon.

At the end of each question in the Sample Test and Additional Test Questions sections, a letter and number will be used as a reference back to this section for additional study. Note the following example: **C.13.**

C. Fuel, Air Induction, and Exhaust System Diagnosis and Repair (14 Questions)

Task C.13 Inspect, service, and replace exhaust manifold, exhaust pipes, mufflers, resonators, catalytic converters, tail pipes, and heat shields.

Example:

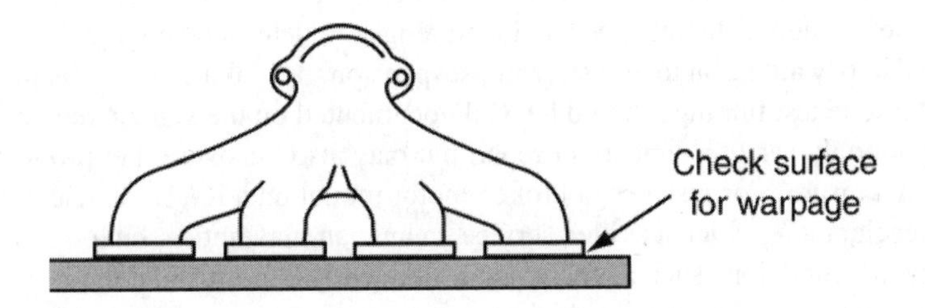

1. To check an exhaust manifold for warpage between ports, as shown in the figure, Technician A says that only a straightedge is needed. Technician B says a straightedge and a flashlight should be used. Who is right?

A. Technician A only
B. Technician B only
C. Both A and B
D. Neither A nor B (C.13)

Analysis:

Question #1
Answer A is wrong. A straightedge should be used with feeler gauges.
Answer B is wrong. A straightedge should be used with feeler gauges.
Answer C is wrong. Neither Technician A nor B are correct.

Answer D is correct. Neither Technician A nor B are correct.

Task List and Overview

A. General Engine Diagnosis (10 Questions)

Task A.1 Verify driver's complaint, perform visual inspection, and/or road test vehicle; determine needed action.

When a customer brings their vehicle to the shop for repair, they have a concern or problem in their mind but may have difficulty expressing the problem in words. In most repair facilities, it is the service writer's responsibility to help gather information and define the problem so the technician can diagnose it.

Many vehicle concerns can be located with a thorough visual inspection. A leaking fuel-pressure regulator diaphragm may cause hard starting and poor mileage complaints.

Simply removing the fuel-pressure regulator vacuum hose and inspecting it for the presence of fuel can isolate the problem without performing detailed fuel system tests. A visual inspection should include: checking all air intake plumbing for cracks or loose clamps; checking vacuum hoses' condition; routing and checking spark plug wires for misrouted or damaged cables; checking wiring harnesses for proper routing and for damage from chafing brackets or components on the engine; and also checking all fluids for proper level and condition.

A clear and detailed description of the customer's concerns must be obtained either from the customer or the service writer before diagnostics begin. A service technician should review previous service history of the vehicle supplied by the customer or located in the shop's service records to narrow the scope of his testing.

Performing a road test is part of almost all service routines. A road test allows the technician to verify a customer concern as well as to verify a successful repair. When test driving a vehicle, the technician should take note of any unusual noises; look for smoke from the exhaust; operate the vehicle through different speed and load ranges to determine engine and transmission performance; and also pay attention to any steering, suspension, or braking problems that may be evident.

All road test findings should be well documented on the vehicle repair order. A thorough test drive will not only verify a concern or repair but may also uncover other problems of which the customer was not aware. For instance, a broken motor mount on a RWD vehicle may allow the engine to lift on acceleration, which a skilled service technician may notice, but of which the customer could be unaware. Situations such as these occur frequently and illustrate the need for following a careful routine when road testing a vehicle.

Task A.2 Research applicable vehicle and service information, such as engine management system operation, vehicle service history, service precautions, and technical service bulletins.

An important part of diagnosis is understanding the system on which you are working. This applies to engine and powertrain management systems as well as other parts of the vehicle. This includes theory of operation information that is usually found in the manufacturer's vehicle service manuals or aftermarket vehicle service manuals from publishers, such as Mitchell, Motors, or Chilton. These manuals will also provide detailed repair information that is necessary to service vehicles today, such as wiring diagrams, components replacement procedures, adjustment procedures, and specifications and test procedures.

Previous vehicle service information should be gathered from the customer if repairs or service were performed elsewhere, as well as any repairs already performed by the shop that should be kept in a customer file or on a shop management computer. Knowing what repairs have already been performed will help eliminate duplicate services.

Further information is available in the form of manufacturer's technical service bulletins (TSBs) and recall notices. These may be in printed or electronic format and are available from the same sources as mentioned above. In electronic format, the bulletins can be searched with a personal computer by vehicle as well as system or symptom. The bulletin can then be printed or viewed on screen.

The Internet also provides many sources for information gathering. Vehicle and parts manufacturer Web sites are available and provide access to service information or the ability to purchase manuals and training material on-line.

Some hand-held equipment available today has diagnostic information imbedded in the tools' software for easy technician access. This form of information access is very useful and well suited to scan tools and labscopes or graphing multimeters. Lastly, vehicle repair diagnostic hotline services are available and can be very helpful resources when working on unfamiliar vehicles or complex problems. Various sources such as parts suppliers, tool manufacturers, and private/independent companies provide these services.

Task A.3 Diagnose noises and/or vibration problems related to engine performance; determine needed action.

Engine defects such as worn pistons and cylinder walls, worn rings, loose piston pins, worn crankshaft bearings, a cracked flywheel or flex plate, worn camshaft lobes, and loose or worn valve train components usually produce their own identifiable noises or vibrations. Identifying when the noises and vibrations occur can be helpful in determining the faulty component. A stethoscope would be useful in determining the location of a noise.

Attention must be paid to the relation of the noise to engine speed as well as engine temperature to help pinpoint the source of the noise. For example, a problem such as piston slap may be very noticeable with a cold engine but almost disappear when the engine reaches operating temperature; while a worn connecting rod bearing gets louder as the engine warms up and the oil thins out.

Task A.4 Diagnose the cause of unusual exhaust color, odor, and sound; determine needed action.

If the engine is operating normally, the exhaust should be colorless. A small swirl of white vapor from the tailpipe is normal in cold weather. This is vapor moisture in the exhaust and is a byproduct of combustion. If the exhaust is blue, small amounts of oil are entering the combustion chamber. If the exhaust is black, the air-fuel mixture is too rich. If the exhaust is gray, coolant may be leaking into the combustion chambers.

Task A.5 Perform engine manifold vacuum or pressure tests; determine needed action.

When a vacuum gauge is connected to the intake manifold, the gauge should show a steady reading between 17 and 22 in. Hg (44.8 and 27.6 kPa absolute) with the engine idling at sea level. Manifold vacuum readings will decrease approximately 1 in. Hg for every 1,000 feet increase in elevation above sea level. A low steady reading indicates late ignition timing. Burned or leaking valves cause a vacuum gauge fluctuation between 12 and 18 in. Hg. When the engine is accelerated and held at a steady 2,500 rpm and the vacuum reading slowly drops to a very low reading, an exhaust system restriction is indicated.

When performing vacuum tests, the technician should keep in mind the effect of the valve train on the production of vacuum. If the valve timing is not correct, the engine will not perform as designed and lower vacuum readings will result. Additionally, incorrect valve adjustment can reduce engine efficiency. Valve adjustment that is too tight (insufficient clearance) can cause a valve to hold open reducing compression and cause the valve to burn. Incorrect valve timing will change when the valves are open and closed and can greatly affect the manifold vacuum. Late valve timing from a jumped timing chain or belt will cause very low vacuum readings. If the valve timing is very late, the engine likely will not start.

A cranking vacuum test will show no vacuum or possibly a gauge needle that bounces back and forth across zero, indicating positive pressure in the intake manifold. This occurs because, as the piston begins its upward travel on the compression stroke, the intake valve is still open. Perform cranking vacuum tests during any no-start diagnosis to ensure proper engine sealing. A normal cranking vacuum reading should be between 1 and 5 in. Hg at sea level.

Task A.6 Perform cylinder power balance test; determine needed action.

A cylinder power balance test is performed to ensure that all cylinders are contributing equally. When individual cylinders are disabled, a noticeable revolutions per minute (rpm) drop is measured. By comparing the amount of rpm drop between cylinders, it can be determined which cylinder has a problem. Performing cylinder power balance testing on late-model computer-controlled engines often requires disabling the idle control system and forcing the engine control system into open-loop by disconnecting the oxygen sensor to prevent the computer from compensating for the disabled cylinder.

Many newer engine-control computers allow the technician to perform cylinder balance testing using a scan tool. The computer will disable the fuel injectors one at a time and command a fixed engine speed and open-loop operation during the test. This test is much easier and faster to perform. A cylinder balance test will only identify which cylinder is producing low power; it cannot pinpoint the exact cause. Further pinpoint testing will be necessary to determine if the problem is a compression, ignition, or fuel delivery issue.

Task A.7 Perform cylinder cranking compression test; determine needed action.

The compression test checks the sealing quality of the combustion chamber. If the compression is lower than specified in one or more cylinders, then the valves and rings are suspect. A wet compression test is useful in deciding whether rings or valves are the problem. If compression comes up during the wet compression test, then rings are the most likely problem. Conversely, if compression does not come up, then leaking valves are the likely problem. When performing a compression test, the engine should be at operating temperature and the throttle should be held open in order to get a more accurate reading.

If it is determined that the valves are the cause of compression loss, check the valve clearance and adjust before continuing with repair. In some cases, valve timing could cause compression loss. If this is the case, the sound of the engine while cranking is distinctive and will lead the technician to investigate valve timing.

Task A.8 Perform cylinder leakage/leak-down test; determine needed action.

The cylinder leakage test (leak-down test) can be used to further pinpoint the problem. A regulated amount of air is introduced into the cylinder and the gauge on the tester will indicate the percentage of that pressure which is leaking. A gauge reading of 0 percent indicates no leakage while a reading of 100 percent indicates that the cylinder does not hold compression. If two adjacent cylinders have excess leakage, a head gasket problem is likely. While testing a cylinder with high leakage, you should try to find where the leakage is going. For example, if air is heard escaping from the exhaust, then a leaking exhaust valve is indicated. Air coming from the intake would indicate a leaking intake valve. Leakage could show up in the radiator, indicating a problem with a head gasket or a cracked cylinder head. Air coming from the PCV valve crankcase oil filter would indicate leakage past the rings.

Task A.9 Diagnose engine mechanical, electrical, electronic, fuel, and ignition problems with an oscilloscope, engine analyzer, and/or scan tool and determine needed action.

Today's computerized engine analyzers perform detailed system tests and can produce very accurate diagnostic printouts for the technician or customer. Automated test sequences for the cranking, charging, ignition, fuel, compression, and emission systems are done when a complete test is performed. Most analyzers also have manual pinpoint test capabilities and can be used as stand-alone ignition and lab-grade oscilloscopes. By utilizing all of the test leads of a computerized engine analyzer and running a complete test, the technician can identify most problems that may be present in any of the systems listed above.

For instance, during a cranking test, the analyzer will measure cranking current per cylinder to determine relative compression, check the starting system for cranking speed and normal current draw, monitor the battery condition, check the ignition KV output, and test the fuel delivery by exhaust analysis of hydrocarbons during cranking. All of these measurements are done in 15 seconds of cranking time without the engine even starting. The remaining systems are testing during automated tests that are performed once the engine is started. Determining what is right with the engine prior to performing extensive pinpoint tests to locate a problem is the greatest strength of a computerized engine analyzer. Use the multi-trace labscope function of the analyzer for detailed pinpoint testing of fuel or ignition system components such as oxygen sensors, throttle position sensors, or ignition primary triggering devices.

The hand-held digital storage oscilloscope (DSO) has become a very useful tool for pinpoint testing of components on today's computer-controlled engines. The DSO can be used for signal analysis while test driving a vehicle if the fault is intermittent or if road load is required to create a problem. The DSO coupled with a current probe is a very powerful tool for testing motors and actuators that have winding or coil-like fuel injectors, ignition coils, and solenoids. Fuel pumps or small electric motors can also be accurately tested with a DSO and current probe. Most DSOs have waveform memory, recall capabilities, and can print waveforms to a printer or to special computer programs for saving and compiling a database of good and bad waveforms. These features can also be used for repair verification and customer documentation of tests or repairs.

Answer D is wrong. An excessive spark plug gap causes high resistance.

Task A.10 Prepare and inspect vehicle and analyzer for HC, CO, CO_2, and O_2 exhaust gas analysis; perform tests and interpret exhaust gas readings.

Many states have emission inspection programs that require vehicle owners to maintain their vehicles to certain standards. An emissions analyzer measures tailpipe emissions. Emission analyzers require a warm-up period and certain calibration intervals.

Some items that may be checked with an emissions analyzer include: air-fuel mixture, cylinder misfire, catalytic converter defects, and leaking head gaskets. A four-gas emissions analyzer is capable of measuring hydrocarbons, carbon monoxide, oxygen, and carbon dioxide.

A five-gas analyzer adds the ability to measure oxides of nitrogen (NO_x), and these units are usually portable. Oxides of nitrogen are mostly produced under load, so a means of measuring this pollutant during driving conditions makes a portable gas analyzer necessary. Many states that have enhanced emissions test programs test for NO_x levels, so portable analyzers have become almost necessary in these areas. Sample dilution is a major problem with exhaust gas analysis, and it is important that the technician checks both the vehicle's exhaust system and the analyzer sample probe and hose to verify they are free from leaks. Prior to performing gas analysis, if the vehicle has an air-injection system, disable it to allow for accurate measurements. Any outside air entering the gas analyzer will skew the readings and may cause incorrect diagnosis.

Diagnosing engine or emissions system problems with a gas analyzer requires understanding of what each of the exhaust gases are and how they are produced so that when the levels are incorrect the technician can determine the most likely cause.

Hydrocarbons are basically unburned fuel molecules. Anything that inhibits proper combustion in the engine can raise hydrocarbons. Normal engine out or pre-converter hydrocarbon levels for late-model closed-loop fuel-control vehicles should be less than 300 parts per million. The catalytic converter can lower this level close to zero.

Over-advanced ignition timing, an overly rich or lean air-fuel mixture, or any problem that causes a cylinder to misfire will raise HC levels. Fuel must be drawn into the cylinder to raise HC levels so a cylinder that has a fuel injector that does not function can misfire; but because no fuel is injected, the HC level will not be high. Carbon monoxide is formed in the cylinder when there is an insufficient amount of air in the mixture for the amount of fuel present. Carbon monoxide is a rich indicator. When the air-fuel mixture is leaner than 14.7:1, the amount of CO produced will be very low.

Normal engine out CO levels for late-model, closed-loop, fuel-control vehicles is less than .75 percent. The catalytic converter can lower this amount to almost zero.

Items that may cause high CO levels on a carbureted engine include incorrect carburetor adjustment, high float level, or dirty or plugged air bleeds in the carburetor.

High CO levels on fuel-injected engines can be caused by high fuel pressure, leaking injectors, or a leaking fuel-pressure regulator. Any sensor inputs that have an effect on fuel delivery on closed-loop fuel control vehicles can cause increased CO emissions if they are incorrect or out of calibration. The most common sensor problems include the oxygen sensor, coolant or air intake temperature sensors, manifold absolute pressure sensor, and mass airflow sensor, and throttle position sensor.

Oxygen is not a harmful exhaust by-product, but is measured by gas analyzers to help determine if the mixture is too lean. Oxygen and carbon monoxide levels are equal at the stoichiometric or ideal air-fuel ratio of 14.7:1. When the mixture goes richer, CO will increase; when the mixture goes lean, O_2 will increase. Oxygen is the lean indicator in the exhaust stream. Normal engine out oxygen levels for computer-controlled vehicles is less than 1.5 percent. High O_2 levels can be caused by sample dilution; check the exhaust system and sample hose for leaks before condemning the engine. Any misfire can increase O_2 levels because the air drawn into the engine is not burned in the misfiring cylinder. If the O_2 level is high and the engine runs good, look for sample dilution. If the O_2 level is high and the engine runs rough, check for vacuum leaks or a cylinder misfire.

Carbon dioxide is a desirable exhaust by-product that is a combustion efficiency indicator. A mechanically sound engine should produce a minimum of 12 percent CO_2. Properly operating late-model closed-loop fuel-control vehicles typically show 13.5 to 15.5 percent CO_2 readings. If the other gases are normal and CO_2 does not go above 12 percent, look for a mechanical problem or exhaust sample dilution.

Oxides of nitrogen are formed when nitrogen and oxygen combine in the combustion chamber when temperatures increase above 2,500°F. When oxides of nitrogen and hydrocarbons mix in the atmosphere and are exposed to sunlight, photochemical smog is created. NO_x is controlled in the engine by adding exhaust gas to the incoming air charge to reduce combustion chamber temperatures below 2,500°F. This is the job of the exhaust gas recirculation system. The three-way catalytic converter reduces the NO_x still remaining in the exhaust steam even further. NO_x levels should be below 1,000 ppm when measured with a five-gas analyzer under road load.

When diagnosing engine problems with a four- or five-gas analyzer, the technician must evaluate all of the gases measured to make an accurate decision concerning what may be causing out-of-specification readings. For instance, a high HC reading does not indicate a rich air-fuel mixture because hydrocarbons can increase from a lean misfire as well a rich running condition. By observing the CO and O_2 readings, a technician can determine if the high HC reading is a result of a rich mixture because CO will also be high while O_2 will be very low. If the CO is low and O_2 is high, a lean misfire or cylinder misfire is indicated.

Task A.11 Verify valve adjustment on engines with mechanical or hydraulic lifters.

Valve lifters are either mechanical (solid) or hydraulic. Solid lifters provide for a rigid connection between the camshaft and the valves. Hydraulic valve lifters provide for the same connection but use oil to absorb the shock that results from the movement of the valve train.

Hydraulic lifters are designed to automatically compensate for the effects of engine temperature. Changes in temperature cause the valve train components to expand and contract. Solid lifters require a clearance between the parts of the valve train. This clearance allows for the expansion of the components as the engine gets hot. Periodic adjustment of this clearance must be made. Excessive clearance might cause a clicking noise. This clicking noise is also an indication of the hammering of the valve train parts against one another, which will result in reduced camshaft and lifter life.

Valve lash on some engines is adjusted with an adjusting nut on the valve tip end of the rocker arm. The clearance is checked by inserting a feeler gauge between the valve tip and the adjusting nut. Some OHC engines have an adjustment disc or shim between the cam lobe surface and the lifter or follower. To adjust valve lash, use a special tool and a magnet.

Prior to checking or adjusting valve clearance, consult the vehicle service manual to determine the type of valve train adjustment method used and whether or not the adjustment must be done with the engine cold or warm. Many late-model engines use hydraulic lash adjusters that eliminate the need for periodic adjustment. If the engine requires periodic valve adjustment, follow the exact procedures to prevent incorrect adjustment and possible valve train damage. If valves are adjusted with insufficient clearance, the valve may hold open when the engine is warm. This will cause a loss of compression and valve leakage and burning. If the valve clearance is excessive, the valve will open later and not open as far as it should. This condition causes the valve timing to be retarded and valve overlap and lift are reduced. This will result in lower power contribution from the cylinder.

Task A.12 Verify camshaft timing; determine needed action.

The camshaft and crankshaft must always remain in the same relative position to each other. They must also be in the proper initial relation to each other. This initial position between the shafts is designated by timing marks. To obtain the correct initial relationship of the components, the corresponding marks are aligned during engine assembly. Verification of this relationship is done by rotating the crankshaft to TDC on cylinder #1 and checking the alignment of the timing marks on both shafts.

Problems such as a no-start condition, lack of engine power, or a low vacuum reading on a vacuum gauge should alert the technician to verify correct valve timing relationship. On distributor-equipped engines that have the distributor driven by the camshaft, if the ignition timing is found to be off by more than a couple of degrees, check the camshaft timing to make sure the timing chain or belt have not jumped or been installed incorrectly. To verify correct valve timing on vehicles with a distributor and chain-driven camshaft, rotate the engine until the ignition timing marks are aligned at TDC. Remove the distributor cap. The ignition rotor should be pointing to the #1 spark plug wire terminal.

On engines with a belt-driven camshaft, the timing covers are usually removed to inspect the timing marks on the camshaft and crankshaft, or rubber access plugs may be present to allow inspection. On distributorless ignition engines, the valve cover may have to be removed to inspect valve operation and position for cylinder #1 when the engine is rotated to TDC on the #1 cylinder compression stroke.

Task A.13 Verify engine-operating temperature, check coolant level and condition, perform cooling system pressure test; determine needed repairs.

The cooling system must operate, be inspected, and be serviced as a system. Replacing one damaged part while leaving others dirty or clogged will not increase system efficiency. Service the entire system to ensure good results. Service involves both a visual inspection of the parts and connections and pressure testing. Pressure testing is used to detect internal or external leaks. Pressure testing can also be used to check the condition of the radiator cap. This type of testing involves applying a pressure to the system or cap.

If the system is able to hold the pressure, there are no leaks in the system. If the pressure drops, there is an internal or external leak.

The vehicle's temperature gauge, a shop temperature gauge, or a hand-held pyrometer can verify engine-operating temperature. Check the condition and level of the coolant as part of a preventative maintenance program. The level of the coolant should be at the level specified by the manufacturer. Check the coolant for the presence of engine oil or other contaminants.

Task A.14 Inspect and test, mechanical/electrical fans, fan clutch, fan shroud/ducting, and fan control devices; determine needed repairs.

Mechanical fans can be checked by spinning the fan by hand. A noticeable wobble or any blade that is not in the same plane as the rest of the blades indicates that the fan needs to be replaced. One of the simplest checks of a fan clutch is to look for signs of fluid loss. Oily streaks radiating outward from the hub shaft mean fluid has leaked out past the bearing seal. Most fan clutches offer a slight amount of resistance if turned by hand when the engine is cold. They offer drag when the engine is hot. If the fan freewheels easily when it is hot or cold, replace the clutch.

Electric cooling fans are mounted to the radiator shroud and are not mechanically connected to the engine. An electric motor-driven fan is controlled by either, or both, an engine coolant temperature switch or sensor and an air-conditioning switch. The controls for the electric cooling fan can be easily identified by referring to the wiring schematic for the vehicle.

Task A.15 Repair and interpret electrical schematic diagrams and symbols.

Electrical schematics, also known as electrical diagrams or wiring diagrams, are the road maps technicians use to properly diagnose and understand any electrical circuit on the vehicle. By studying and completely understanding an electrical schematic, many technicians are able to identify electrical problems with no other diagnostic information.

Understanding the symbols used on the schematics is a major key to fully utilizing the diagrams as a diagnostic tool. A technician must be able to identify where the circuit power originates, whether the device is voltage-controlled or ground-controlled, and what other devices are powered from the same source. Identifying terminal pin numbers, wire colors, which fuse(s) feed the circuit and their location, circuit identifying numbers, splices, connectors, and where a circuit is grounded are all quite necessary in nearly every type of electrical operation, and are all possible if a technician has access to the proper wiring schematic.

Most diagnostic procedures are written with the assumption that the technician performing the test is familiar with the location and types of circuits being tested. Having a specific electrical schematic of that circuit in front of the technician, and the technician's ability to gain information from the schematic will ensure that the technician is able to identify what is being tested with the most efficiency.

Many computer or internet-based information systems provide the ability to print out complete schematics or even to zoom in on a specific portion of it and print the zoomed image. The technician can then write notes on the printed copy or even highlight current flow or other pertinent items to focus on.

B. Ignition System Diagnosis and Repair (10 Questions)

Task B.1 Diagnose ignition system-related problems such as no-starting, hard-starting, engine misfire, poor drivability, spark knock, power loss, poor mileage, and emissions problems; determine root cause; determine needed repairs.

Diagnosing engine performance and no-start problems related to the ignition system requires a thorough understanding of the type of ignition system used and the components used in the primary and secondary circuits. Engines will use either a distributor ignition (DI); electronic distributorless ignition using a coil for each pair of spark plugs, called "waste spark" systems (EI); or a coil-on-plug (COP) ignition system. Most ignition systems used since the mid-1980s will incorporate computer control of ignition timing.

No-start diagnosis involves checking for the presence of spark with a spark tester first. If no spark is present at the plugs, the problem must be isolated to either the primary or secondary circuits. Testing the primary circuit with a test light or scope will determine if the coil is being switched on and off. If the coil does not switch on and off, the problem is in the primary circuit and a bad triggering device, module, or wiring is the cause. If the coil is being switched on and off, then the coil secondary windings, a distributor cap and rotor, or secondary wiring is at fault.

Perform further pinpoint testing of ignition system performance with a scope. Testing for available secondary voltage and performing a wet test by spraying water on the secondary wiring will help isolate problems that can cause many different drivability concerns like misfire and hard starting. Consulting specific manufacturer test procedures will be necessary when troubleshooting concerns such as spark knock, poor mileage, or emissions problems due to the many different computer-controlled systems in use.

Defects in the ignition system or incorrect adjustments can cause many common drivability problems. Over-advanced ignition timing will cause pinging or spark knock and increase hydrocarbon emissions; while retarded ignition timing will cause a lack of power and poor mileage. Worn or damaged secondary ignition components can cause engine misfire or rough running complaints and will increase tailpipe emissions.

Secondary voltage leakage from damaged insulation on spark plug wires or cracked spark plug insulators can cause bucking and hesitation complaints that may only occur under road load conditions.

When diagnosing ignition system problems, always consider what caused a failure to occur. For instance, if an ignition module fails, it may be due to increased current flow from a shorted primary winding in the ignition coil. Replacing the ignition module may allow the car to start, but the replacement module will fail prematurely. By performing a more thorough testing routine and checking the ignition coil windings for proper resistance, the root cause of the module failure would be uncovered. Often when dealing with electronic component failures, there is another component problem that will cause a related part to fail. Finding and correcting the root cause of a failure is the difference between a parts changer and a technician.

Task B.2 Interpret ignition system-related diagnostic trouble codes (DTC); determine needed repairs.

DTCs can be retrieved from the powertrain control module (PCM) on most vehicles. The DTCs are displayed in the instrument panel or on a scan tool. The latter is most common. Displayed trouble codes are identified in the service manual along with test procedures necessary to isolate the cause. A DTC usually identifies a problem area, not the exact cause. Performing the pinpoint tests found in the service manual should determine the exact problem causing the code to set. Some vehicles have more in-depth ignition system trouble code capabilities and can set DTCs for primary ignition components like modules or triggering devices, while other systems may only set codes for spark timing control problems.

Task B.3 Inspect, test, repair, or replace ignition primary circuit wiring and components.

Ignition system primary wiring faults will most often be seen as bad connections at terminals and connectors. Probing wiring through insulation to make electrical tests will damage the insulation and can lead to wiring corrosion and high resistance especially in climates where road salt is used in the winter. Test primary wiring for unwanted resistance and current-carrying capacity by performing available voltage and voltage drop tests or measuring resistance with an ohmmeter. When replacing primary wiring, use the correct gauge wire. Solder connections and seal the repair with heat-shrink tubing to provide adequate weatherproofing.

Test primary components such as the ignition switch or ballast resistor, if used, with a digital multimeter (DMM). Check both power and ground circuits to insure proper primary circuit current flow and operation.

Task B.4 Inspect, test, and service distributor.

Check the distributor to confirm that it is functioning correctly mechanically. Evaluate bushings, centrifugal advance, and body condition and repair as required. Distributor testers are available that can test the distributor dynamically. This is particularly useful in checking the effect of shaft play on the performance of the distributor.

Before installing and timing the distributor, ensure that the engine is at top dead-center (TDC) on the compression stroke of the specified cylinder. In most cases, this is #1 cylinder, but some manufacturers specify a different cylinder for particular engines. Position the distributor rotor so it will point toward the specified cap terminal when the distributor cap is installed.

Task B.5 Inspect, test, service, repair, or replace ignition system secondary circuit wiring and components.

Visible inspect the ignition system secondary circuit wiring for chafing against metal brackets or exhaust manifolds that could cause arcing as well as proper routing. Misrouted wires can increase the possibility of cylinder crossfire that can cause extreme engine damage. Check wires and spark plug boots for coolant or oil-soaked conditions and replace them if these conditions exist. Check spark plug wire terminals for signs of corrosion or arcing at both ends of the wire. Check wires for proper resistance with an ohmmeter; specifications are listed in the service manual and are usually expressed in ohms per foot.

Inspect distributor caps and rotors for burned terminals and cracks as well as secondary voltage arcing damage. Use an ignition oscilloscope for leakage in secondary wiring. Spray a fine mist of water on the wires to simulate damp conditions that cause many secondary ignition problems. Worn spark plug gaps or excessive distributor cap-to-rotor air gaps are easily tested with an ignition scope.

Task B.6 Inspect, test, and replace ignition coil(s)

Inspect the ignition coil for cracks or any evidence of leakage in the coil tower. Check the coil container for oil leaks. If oil is leaking from the coil, air space is present, allowing condensation to form internally. Condensation in an ignition coil causes voltage leaks and engine misfire. When testing the coil with an ohmmeter, most primary windings have a resistance of 0.5 ohms to 2 ohms and secondary windings have a resistance of 8,000 ohms to 20,000 ohms. The maximum coil output can be tested with an engine analyzer. Always refer to the manufacturer's specifications.

Task B.7 Check and adjust, if necessary, ignition system timing and timing advance/retard.

Ignition timing specifications and instructions are included on the under-hood emissions label. A timing light is connected to the #1 cylinder spark plug wire and to the vehicle battery. The vehicle must be at the specified revolutions per minute (rpm) when the timing light is aimed at the timing indicators. Observe the timing marks. If the timing marks are not at the specified location, rotate the distributor until the mark is at the specified location and tighten down the distributor.

Computer-controlled ignition systems will require disabling the timing advance system when checking base timing. This is usually done by unplugging a connector or installing a jumper wire between the correct terminals of a diagnostic connector. Once the computer is disabled, the timing can be set correctly. After setting base timing, the engine speed should be increased to verify that timing advance is being supplied by the computer. Consult the service manual when checking timing advance because some systems may not advance timing when the vehicle is in park.

Task B.8 Inspect, test, and replace ignition system pickup sensor or triggering devices.

There are several different styles of primary triggering devices used on ignition systems. Magnetic pulse generators, Hall-effect switches and optical pickups are the most common. When testing magnetic pickups, use an ohmmeter to test for proper resistance of the coil and for a grounded or open coil. Connect the ohmmeter leads across the coil leads to check for proper resistance or an open coil. Connect one meter lead to a pickup coil lead and the other to ground to test for a grounded coil. Other tests that can be performed on magnetic pickups include testing AC voltage output with a DMM and signal waveform measurements with a scope. Consult the service manual for specifications for these tests.

Hall-effect switches and optical pickups require a voltage feed and ground to operate properly. Once proper feed voltage and a good ground are verified, test the signal line. These pickups are often supplied a reference voltage from the ignition control module or powertrain control computer, usually between 5 to 10 volts. A labscope is the best method for testing the signal from these types of pickups.

Some Hall-effect and optical pickups may have four wires. In this case, there will be two signal lines on the labscope trace. Often this will be a low and high data rate signal. Some optical pickups generate a square wave signal output for every degree of camshaft rotation, so the signal created will be a very high frequency. While a DMM that measures frequency can determine if a signal is being generated, a labscope provides a much better picture of signal integrity. The square wave signal from a Hall-effect or optical pickup should reach 90 percent of reference voltage when the signal is high and pull down to within 10 percent of ground potential when the signal is low.

Task B.9 Inspect, test, and/or replace ignition control module (ICM)/power train control module (PCM).

Many ignition module testers are available from vehicle and test equipment manufacturers. These testers check the module's capability of switching the primary ignition circuit on and off. On some testers, a green light is illuminated if the module is satisfactory; the light remains off when the module is defective. Always follow the manufacturer's recommended procedure. If the module tests satisfactory, then perform circuit tests to confirm that the wiring of the circuit is serviceable and the proper signals are going to the correct designations.

The ignition module removal and replacement procedure varies, depending on the ignition system. Always follow the manufacturer's recommended replacement procedure.

Some ignition modules require the use of dielectric silicone grease for heat dissipation through the mounting surface. Clean the mounting surface and apply a light coat of silicone grease to the module prior to installing the module. If silicone is not used, heat generated by the modules transistor will not be properly dissipated, and early failure of the module may occur.

Replacement of a powertrain control module should be done only after power and ground circuits to the PCM have been tested and all PCM outputs have been checked for proper current draw to prevent repeated failure. Be careful to ground yourself to dissipate any static electrical charges from your body prior to handling the replacement PCM. This will prevent damage to sensitive circuit boards from any static electrical discharge.

C. Fuel, Air Induction, and Exhaust System Diagnosis and Repair (11 Questions)

Task C.1 Diagnose fuel system-related problems, including hot or cold no-starting, hard-starting, poor drivability, incorrect idle speed, poor idle, flooding, hesitation, surging, engine misfire, power loss, stalling, poor mileage, and emission problems, determine root cause; determine needed action.

The first step in diagnosing fuel system-related problems is to identify the type of system used on the vehicle and review its theory of operation. Obtain a clear description of the problem and verify the complaint.

Once the problem is verified, check for any relative service bulletins before proceeding with detailed testing. Once the complaint is isolated to the fuel system, perform pinpoint tests of the fuel delivery system to determine the exact cause of the problem.

Always seek to determine the root cause of a failure to prevent repeated problems. An example is a fuel pump that fails due to debris in the fuel tank. Replacing the pump may correct the drivability problem, but the root cause of the fuel pump failure was dirt in the fuel tank. If the contamination is not found and removed from the tank, the new fuel pump will fail prematurely.

Problems related to a lack of fuel delivery include power loss, hesitation, surging, and hard starting. Excessive fuel delivery from items such as high fuel pressure, leaking injectors, or a leaking or stuck fuel-pressure regulator can cause a no- or hard-start condition; flooding; engine misfire and poor idle quality; poor fuel economy; and excessive exhaust emissions. Any restrictions in the air inlet or exhaust system can cause reduced engine power, poor mileage and surging or hesitation complaints.

Task C.2 Interpret fuel or induction system-related diagnostic trouble codes (DTCs); determine needed repairs.

Fuel system-related DTCs can be retrieved from nearly all computer-controlled vehicles. Older systems support both flash code and scan tool diagnostics; while 1996 and newer vehicles use primarily scan tool code retrieval. Fuel system trouble codes rarely point to specific components but most often are set because the computer can no longer properly control the air-fuel mixture at ideal levels. Once trouble codes are obtained, consult the service manual to perform the tests necessary to pinpoint the exact cause of the problem.

Often diagnostic trouble codes point to a system, not a component, and the technician must consider what the computer sees in order to set a trouble code. An example would be a lean exhaust trouble code. If the oxygen sensor fails, it no longer produces an output voltage. The computer sees the low-signal voltage and interprets this as a lean exhaust signal. The computer will increase fuel delivery in an attempt to generate a high-voltage signal from the oxygen sensor. The engine can be running very rich and an exhaust gas analysis will confirm this, yet the computer trouble code is set for a lean exhaust signal because of the failed oxygen sensor.

Task C.3 Inspect fuel tank, filler neck, and gas cap; inspect and replace fuel lines, fittings, and hoses; check fuel for contaminants and quality.

Inspect the fuel tank for leaks; road damage; corrosion; rust; loose, damaged, or defective seams; loose mounting bolts; and damaged mounting straps. Leaks in the fuel tank, lines, or filter may cause gasoline odor in and around the vehicle, especially during low speed driving and idling. In most cases, the fuel tank must be removed for service.

Inspect nylon fuel pipes for leaks, nicks, scratches, cuts, kinks, melting, and loose fittings. If these fuel pipes are kinked or damaged in any way, replace them. Nylon fuel pipes provide a certain amount of flexibility and can be formed around gradual curves under the vehicle. Do not force a nylon fuel pipe into a sharp bend because this action may kink the pipe and restrict the flow of fuel.

Obtain a sample of the fuel and examine for dirt or other contaminants. Use a commercially available alcohol tester to determine the percentage of alcohol present in the fuel. Generally fuel-injection systems will tolerate only a limited amount of alcohol in the fuel. Consult workshop manual for details.

Task C.4 Inspect, test, and replace fuel pump and/or fuel pump assembly; inspect, service, and replace fuel filters.

When testing mechanical and/or electric fuel pumps, pressure and volume tests apply. Fuel must be available to the engine at the correct pressure and adequate volume for proper operation. It is possible to have the correct pressure with little or no volume. Fuel volume is the amount of fuel delivered over a specified period of time. The correct amount is specified by the manufacturer and is generally about 1 pint in 30 seconds.

Exercise caution when performing volume testing because fuel is discharged into an open container, which creates a risk of fire. Test mechanical fuel pumps for both vacuum and pressure by alternately connecting a vacuum/pressure gauge first to the inlet fitting with the outlet removed and cranking the engine, then reconnecting the inlet line and starting the engine with the gauge teed into the outlet line to the carburetor. Electric fuel pumps can be tested for proper pressure by connecting a fuel-pressure gauge to the Schrader valve test port on the fuel rail. If the vehicle does not have a test port on the fuel rail, the gauge will need to be teed into the system using the correct adaptors.

Low pressure can result from a restricted fuel line or filter, a defective pump, or a lack of good voltage supply and ground connections in the pump's electrical circuit. If fuel pressure is higher than specified, a bad fuel-pressure regulator or a restricted fuel return line could be the cause. Most fuel-injection systems should hold pressure after the engine is turned off. A bad fuel pump outlet check valve, a bad fuel-pressure regulator, or a leaking fuel injector can cause a pressure leak-down. Follow the manufacturer's test procedures to isolate the failed component.

Replace fuel filters at the manufacturer's specified intervals or more often, if the fuel quality in the area is poor or if the vehicle has had fuel contamination problems.

Task C.5 Inspect and test fuel pump electrical control circuits and components; determine needed repairs.

Electric fuel pumps are supplied voltage through a variety of different control circuits and components depending upon the manufacturer. Most commonly, the computer will control a relay that supplies current to the pump. The computer will energize this relay when the key is first turned on for two seconds to prime the system. If the engine is not started, the computer shuts off the relay after the prime period. If the engine were to stall, the computer will again turn off the relay. Once the computer receives an rpm signal indicating the engine is running, the fuel pump relay will remain energized. Many cars also use an inertia switch in the feed line to the pump. In case of an accident, the inertia switch will open the circuit preventing fuel pump operation and reducing the risk of a fire.

On some import vehicles, the fuel pump control circuit goes through contacts in the vane style airflow meter. Once again with this system, if the engine does not stay running the fuel pump will not operate.

If there is a fuel delivery problem or no-start condition because of a lack of fuel, obtain wiring schematics and determine the type of control circuit used to power the fuel pump. With this information, perform the necessary electrical tests on the components in the system to isolate the problem.

Task C.6 Inspect, test, and repair or replace fuel-pressure regulation system and components of fuel-injection systems; perform fuel pressure/volume test.

Test the fuel-pressure regulator for leakage through the diaphragm by removing the vacuum hose with the engine running to see if fuel drips from the vacuum nipple. Fuel pressure should reach the pressure specified by the manufacturer with the vacuum line disconnected and lower by about 10 psi when the vacuum line is plugged back onto the regulator. Excessive fuel pressure indicates a stuck closed regulator or restricted fuel return line. Low pressure can be caused by a weak fuel pump, restricted fuel filter, or stuck open regulator. If the pressure increases to normal when the return line is restricted, replace the regulator. If fuel pressure drops when the engine is turned off, a leak is indicated in the fuel system. Alternately clamping off the fuel feed and return lines will isolate the location of the leak. If the pressure drop stops when the return line is pinched, the fuel-pressure regulator is leaking and should be replaced. If the pressure still drops, a leak through the fuel pump or injectors is the problem. Clamping the fuel feed line will eliminate the fuel pump as the source of the pressure drop if the system holds pressure; then the injectors will need service. Do not clamp plastic fuel line as this will cause permanent damage. Install rubber test lines to perform this test if the vehicle has plastic fuel lines.

Task C.7 Inspect, remove, service or replace throttle assembly; make related adjustments.

After many miles of operation, an accumulation of gum and carbon deposits may occur around the throttle area on throttle body injected (TBI), multiport fuel-injected (MFI), and sequential fuel-injected (SFI) systems. This can cause rough idle and stalling problems. Use throttle body cleaner to spray around the throttle area without removing and disassembling the throttle body. If this cleaning does not remove the deposits, remove the throttle body according to the manufacturer's recommendations, disassemble, and place it in an approved cleaning solution. Remove the throttle position sensor (TPS), idle air control (IAC), fuel injector, pressure regulator, and seals prior to placing the throttle body into the cleaning solution.

Since MFI and SFI systems do not have the pressure regulator or fuel injectors in the throttle body, these items need not be removed. After cleaning, check the closed throttle position or minimum air rate and adjust, if necessary, along with setting the throttle position sensor voltage. Consult the service manual for the correct procedures to follow when making these adjustments.

Task C.8 Inspect, test, clean, and replace fuel injectors.

Fuel injectors can be tested on the car by performing an injector balance test. While monitoring fuel pressure, each injector is fired one at a time with a special tool designed to open the injector an exact amount of time. When the injector is triggered, the fuel pressure in the rail will drop. No pressure drop indicates a plugged injector or open coil.

An injector with a low-pressure drop indicates a dirty or restricted injector. An injector with too great a pressure drop indicates a leaking or rich injector. A pressure difference greater than 1.5 psi above or below the average is considered a problem that requires service. Fuel injectors should also be tested with a labscope to observe their electrical integrity. Both voltage and current waveforms can be observed with a labscope.

Tool manufacturers market a variety of fuel-injector cleaning equipment. A solution of fuel-injector cleaner is mixed with unleaded gasoline. Shop air pressure provides system operating pressure. The vehicle's fuel pump must be disabled to prevent fuel from being forced into the fuel rail. Plug the fuel return line to keep the fuel-injector cleaner solution from entering the fuel tank. After the fuel injectors have been cleaned, reset the adaptive memory.

If the injectors have been removed for cleaning, check the spray pattern. An even cone-shaped pattern without thready or dripping discharge should be present. If the proper spray pattern cannot be achieved, replace the injector.

Task C.9 Inspect, service, and repair or replace air filtration system components.

If a vehicle is operated continually in dusty conditions, air filter replacement may be necessary at more frequent intervals. A damaged air filter can cause increased wear on cylinder walls, pistons, and piston rings. If the air filter is restricted with dirt, it restricts the flow of air into the intake manifold, and this increases fuel consumption.

Task C.10 Inspect throttle assembly, air induction system, intake manifold, and gaskets for vacuum leaks and/or unmetered air.

Intake manifold vacuum leaks can cause rough idle and stalling complaints or incorrect idle speeds, and may cause trouble codes to be logged for idle speed or fuel trim errors. Vacuum leaks can be located by flowing propane around suspected areas to see if the idle is affected. A smoke machine is a popular tool for locating vacuum leaks and can be used without running the engine. Not only will intake leaks be located with a smoke machine, but any vacuum accessory connected to the intake manifold will also be checked and pinpointed if leaking. A careful visual inspection should be performed on all air induction hoses to locate any cracks or loose clamps. This is especially important on mass airflow-equipped engines because unmetered air drawn into the engine behind the airflow meter will cause lean running conditions and a lack of power or hesitation complaints.

Task C.11 Check and/or adjust idle speed where applicable.

The manufacturer's procedures must be followed when setting idle speeds on modern vehicles. Carbureted engines may have multiple idle control settings if solenoids or idle speed control motors are used. Fuel-injected engines usually require a minimum throttle setting, called minimum air rate adjustment, to allow the computer to properly control idle speed. Some systems need to have an idle re-learn procedure performed if the battery is disconnected or the throttle plate is cleaned or adjusted. Some other fuel-injected engines use an air by-pass screw to set idle speed. Make sure the engine is operating under the proper conditions to set idle speed such as a specified gear position, or if any diagnostic connectors must be jumped or unplugged.

Task C.12 Remove, clean, inspect, test, and repair or replace fuel system vacuum and electrical components and connections.

Fuel system vacuum and electrical components include the fuel-pressure regulator and any vacuum controls if used, vacuum-operated throttle positioner, fuel pump relay, inertia switch, two-speed fuel pump resistor, and electronic fuel pump power modules. A visual inspection will uncover damaged vacuum lines and proper routing can be checked against the under-hood emissions label. Visually check all electrical connections for terminal seating as well as damaged, chafed, or corroded wires or connections. Basic electrical testing with a test light and DMM can determine problems with fuel system electrical components. Consult the service manual to identify which components are used and what if any special test procedures may be required.

Task C.13 Inspect, service, and replace exhaust manifold, exhaust pipes, mufflers, resonators, catalytic converters, tail pipes, and heat shields.

Remove the exhaust pipe bolts at the manifold flange, and disconnect any other components in the manifold, such as the O_2 sensor. Remove the bolts retaining the manifold to the cylinder head, and lift the manifold from the engine compartment.

Remove the manifold heat shield. Thoroughly clean the manifold and cylinder head mating surfaces. Measure the exhaust manifold surface for warping with a straightedge and feeler gauge in three locations on the manifold surface. Examine the manifold carefully for any cracks or broken flanges.

Follow the exhaust system from manifold to tail pipe end. Ensure that all hangers are present and installed correctly. The exhaust system is designed to be suspended from these hangers; loosen joints and realign if any of the hangers are in tension. Examine all pipes, mufflers, and resonators to ensure that they are securely connected and gas tight.

Task C.14 Test for exhaust system restriction; determine needed action.

A restricted exhaust pipe, catalytic converter, or muffler may cause excessive exhaust back pressure. If the exhaust backpressure is excessive, engine power and maximum vehicle speed are reduced. Even a partial restriction will reduce performance and fuel mileage. A low-range pressure gauge or compound vacuum gauge can be connected into the exhaust system to measure backpressure directly. Backpressure test adaptors are available that screw into the oxygen sensor hole and provide a hose nipple for connection of a compound vacuum gauge. Another adaptor looks like a rivet that is installed into a hole drilled into the exhaust system front pipe and allows a gauge to be screwed into the inside threads of the adaptor. Generally, exhaust backpressure should measure less than 3 psi maximum when engine speed is held around 2,500 rpm.

Another method of testing for exhaust restrictions is to measure intake manifold vacuum. Normal vacuum at idle should be between 16 to 21 in. Hg (48.3 to 31 kPa absolute). When the engine is accelerated to 2,500 rpm and held, the vacuum reading will drop momentarily and then stabilize equal to or greater than the idle reading. A vacuum reading that drops very low or to zero indicates a restricted exhaust system.

Task C.15 Inspect, test, clean, and repair or replace turbocharger or supercharger and system components.

Both a turbocharger and supercharger increase the amount of air delivered to an engine's cylinders by increasing the amount of pressure at which the air is delivered. A turbocharger is driven by the velocity and heat of the exhaust leaving the engine. Intake air pressure is increased by a compressor in the turbocharger unit. The faster the compressor turns, the more the air is boosted. The speed of the compressor is determined by the load and speed on the engine. To control the boost and therefore prevent overboost, turbochargers are equipped with a wastegate that controls the amount of exhaust gas at the turbocharger.

A supercharger is driven by the engine's crankshaft via a drive belt. The speed of the compressor or supercharger is directly related to the speed of the engine. The pressure boost for either system can be measured with a pressure gauge connected to the intake manifold. During a road test, the pressure can be observed during a variety of speed and load conditions. Recording the condition and the resulting pressure can lead to a thorough evaluation of the turbocharger or supercharger system. Both a supercharger and turbocharger are non-serviceable items. If there is a problem with either unit, it is replaced. Only the control circuits of these systems can be serviced.

D. Emissions Control Systems Diagnosis and Repair (including OBD II) (9 Questions)

Task D.1 Positive Crankcase Ventilation (PCV) (1 Question)

Task D.1.1 Test and diagnose emissions or drivability problems caused by positive crankcase ventilation (PCV) system.

If the PCV valve is stuck in the open position on a carbureted engine, excessive airflow through the valve causes a lean air-fuel ratio and possible rough idle operation or engine stalling. On a fuel-injected engine, a stuck open PCV valve can cause high idle speed complaints. When the PCV valve or hose is restricted, excessive crankcase pressure forces engine blowby gases through the clean air hose and filter into the air cleaner housing.

Oil accumulation and crankcase blowby gases may also be found in the air cleaner housing on high mileage engines because excessive engine blowby pressure from worn piston rings and cylinders will create more crankcase pressure than the PCV system can handle. Internal engine repairs are the only means of correcting this problem. Flow meter style testers are available to test for excessive crankcase blowby.

Task D.1.2 Inspect, service, and replace positive crankcase ventilation (PCV) filter/breather cap, valve, tubes, orifices/metering device, and hoses.

A thorough examination of the PVC system is relatively easy. After performing the recommended diagnostics, visually inspect the cap, tubes, and hoses for kinks, cuts, or other damage. Disassemble the PVC system to isolate the cause of the restriction. Shake the PCV valve next to your ear and listen for the tapered valve rattling inside the housing. If no rattle is heard, replace the PCV valve.

PCV diagnostic recommendations differ from manufacturer to manufacturer. Some recommend removing the PCV valve and hose from the rocker cover. Connect a length of hose to the inlet side of the PCV valve, and blow air through the valve with your mouth while holding your finger near the valve outlet. Air should pass freely through the valve. If not, replace the valve. Connect a length of hose to the outlet side of the PCV valve and try to blow back through the valve. If air passes easily through the valve, it should be replaced. Other manufacturers recommend disconnecting one end of the PCV valve and placing a finger over it with the engine idling. When there is no vacuum at the PCV valve, part of the system is restricted.

Task D.2 Exhaust Gas Recirculation (3 Questions)

Task D.2.1 Test and diagnose drivability problems caused by the exhaust gas recirculation (EGR) system.

The EGR valve should open once the engine is warm and run above idle or under road load conditions. If the EGR valve remains open at idle and low engine speed, rough idle and stalling can occur as well as engine surging during low-speed driving conditions.

When this problem is present, the engine may also hesitate on low-speed acceleration or stall after deceleration or after a cold start. If the EGR valve does not open, engine detonation can occur and emissions levels will increase. EGR system problems can affect emissions levels differently depending on the problem. A stuck open EGR valve will create a density misfire condition in which the air-fuel mixture is diluted with exhaust gas causing misfire in the combustion chamber. This problem will increase hydrocarbon emissions and raise oxygen levels. An EGR valve that does not open will cause an increase in oxides of nitrogen emissions.

By introducing exhaust gas into the engine during acceleration and cruise conditions, the inert exhaust gas helps reduce peak cylinder combustion temperatures below 2,500°F.

When temperatures rise above 2,500°F, oxygen combines with nitrogen to form oxides of nitrogen (NO_x). Oxides of nitrogen are a main contributor to photochemical smog and must be controlled to manageable levels inside the engine because the three-way catalytic converter is not very efficient at reducing NO_x. If the EGR valve does not open, combustion chamber temperatures will rise and NO_x production will increase far beyond what the catalytic converter can control.

Task D.2.2 Interpret exhaust gas recirculation (EGR)-related diagnostic trouble codes (DTCs); determine needed repairs.

DTCs for the exhaust gas recirculation system will usually be set for one of three reasons: a control circuit fault, a no- or low-flow condition, or an excessive-flow or flow-when-not- commanded condition. The DTC should identify the problem area but not necessarily the component at fault. Once the DTC is retrieved, consult the service manual to determine the necessary tests to be performed to isolate the cause of the code.

Task D.2.3 Inspect, test, service, and replace components of the EGR system, including EGR valve, tubing, exhaust passages, vacuum/pressure controls, filters, hoses, electrical/electronic sensors, controls, solenoids, and wiring of exhaust gas recirculation (EGR) systems.

The first step in diagnosing any EGR system is to check all of the system's vacuum and electrical connectors. In many systems, the PCM uses inputs from various sensors to operate the EGR valve. Improper EGR operation may be caused by a defect in one or more of the sensors. DTCs should be retrieved and the cause corrected before any further diagnostics are completed.

There are both vacuum-operated and electronic EGR valves in use. Test vacuum-operated EGR valves for proper vacuum supply to the valve. Use a vacuum pump to apply vacuum to the valve to test the diaphragm and see that the valve opens. Some valves may require that the engine be running and off idle to operate the internal backpressure transducer so that the valve will hold vacuum. A noticeable change in engine speed and idle quality should be observed. This confirms the EGR passageways are not plugged. No change in engine operation will require removing the valve and cleaning the passages. A scan tool may be necessary to test electronic EGR valves properly. Vacuum-bleed filters can be used on some systems and may require periodic replacement if they become clogged.

Often EGR problems are caused by faulty EGR controls, such as the EGR vacuum regulator (EVR). This regulator can be checked with a scan tool or an ohmmeter. Connect the meter across the terminals of the EVR. An infinite reading indicates there is an open inside the EVR; whereas a low resistance reading means the EVR's coil is shorted internally. The coil should also be checked for shorts to ground. To do this, connect the meter at one of the EVR terminals and the other to the case. The reading should be infinite. If there is any measured resistance, the unit is shorted.

Other EGR control components can also be checked on the scan tool or with a DMM. Refer to the appropriate service manual for the exact procedures and the desired specifications.

Task D.3 Secondary Air Injection (AIR) and Catalytic Converter (2 Questions)

Task D.3.1 Test and diagnose emissions or drivability problems caused by the secondary air injection or catalytic converter systems.

Catalytic converters have been in use since 1975. They are placed in the exhaust system shortly after the engine and are designed to clean the exhaust of excess pollutants through a chemical reaction. There are three basic types of catalytic converters:

Two-way converters are used primarily on pre-1980 vehicles and control unburned hydrocarbons (HC) and carbon monoxide (CO).

Three-way converters (TWC) control HC, CO, and oxides of nitrogen (NO_x). These converters are used on 1981 and later vehicles with computerized engine controls.

Three-way plus oxidation converters are used on 1980 and later vehicles with computerized engine controls and air injection.

The most common reasons for converter failure are overheating and contamination from oil burning or leaded fuel. An engine misfire or extremely rich air-fuel mixture can allow unburned fuel to enter the converter, which can cause excessive heat and converter failure.

Secondary air injection is used to reduce hydrocarbon (HC) and carbon monoxide (CO) emissions by oxidizing these pollutants in the exhaust manifold or catalytic converter. On some vehicles, outside air is injected into the exhaust manifold or converter by a belt-driven or electric air pump or by a pulse air-injection system. The air is routed through hoses and pipes by control valves and one-way check valves during certain engine operating conditions and mixed with exhaust gases as they leave the engine. Check valves installed at the exhaust manifold and catalytic converter air supply pipes prevent exhaust gases from flowing back into the control valves or air pump and damaging these components.

Air-injection control valves consist of a diverter valve (used to dump air pump output to atmosphere during deceleration to prevent backfiring) and switching valves that send air pump output to either the catalytic converter or exhaust manifold, depending on engine operating conditions.

On closed-loop feedback fuel-control systems, the air pump output is directed to the exhaust manifold after starting and during warmup. This allows faster warmup of the oxygen sensor and catalytic converter. Once closed-loop fuel control is entered, the air pump output is switched to the catalytic converter to provide additional oxygen for the rear bed or oxidation bed of the converter. If a dual-bed converter is not used, the air pump output is diverted to atmosphere. If the air pump fails or the hoses are disconnected, tailpipe emissions will increase. If air is not diverted away from the exhaust during deceleration conditions, exhaust backfiring may result.

Another type of secondary air injection is known as a pulse air-injection system. On this system, outside air is drawn into the exhaust manifold by negative pressure pulses created as the exhaust is pushed out of a cylinder by the piston. This system requires no power from the engine to run a pump as in the belt-driven varieties. A reed valve that is sensitive to the negative pressure pulses opens to allow airflow into the exhaust but closes when positive exhaust pressure is present to prevent hot exhaust gases from backing up into the fresh air supply line that is usually connected to the air cleaner housing. This reed valve may also be known as an aspirator valve but the function is the same. Pulse air-injection systems are most efficient at low engine speeds. At higher speeds, the exhaust pulses occur too rapidly and very little air is drawn into the exhaust system.

Task D.3.2 Interpret secondary air-injection system-related diagnostic trouble codes (DTCs); determine needed repairs.

Some vehicles can set DTCs for secondary air-injection system problems, such as control circuit fault codes, and airflow switching problems (such as air not being delivered to the exhaust manifold when commanded). Other vehicles may have no computer diagnostic capabilities for the secondary air-injection system. Consult the service manual for specific models. Catalytic converter efficiency is monitored on OBD II-compliant vehicles and will set DTCs if the converter becomes degraded. Follow manufacturer-specific test routines if a catalyst DTC is set.

Task D.3.3 Inspect, test, service, and replace mechanical components and electrical/electronically operated components and circuits of secondary air-injection systems.

Check all hoses and pipes in the system for looseness and rusted or burned conditions. Burned or melted air-injection hoses or valves indicate leaking check valves.

Inspect the check valves and replace them if they show signs of leakage. With the engine idling, listen for noises from the pump (if equipped). Check the air pump drive belt; adjust it if loose, or replace it if worn or damaged. Check for adequate airflow from the pump and test for airflow to the exhaust manifold during engine warmup and for flow to the catalytic converter when the fuel system enters closed loop.

A properly operating air pump should raise tailpipe oxygen readings above 2 percent and often show levels as high as 3 to 8 percent. Airflow should also divert to the air cleaner or atmosphere when the engine is decelerated rapidly or during high rpm operation. This is the function of the air pump diverter valve. The air-injection switching valve controls airflow to the exhaust manifolds or catalytic converter depending upon engine operating conditions. On some computer-controlled vehicles, the PCM can control the diverter and switching valves according to engine operating conditions. The computer commands to the air-injection system valves can be checked on a scan tool. A scan tool may allow testing on some electric air pumps by means of allowing the technician to turn the air pump on and off. Check the vehicle service manual to identify system components and determine proper test procedures.

Task D.3.4 Inspect catalytic converter. Interpret catalytic converter-related diagnostic trouble codes (DTCs); determine needed repairs.

If the catalytic converter rattles when tapped with a soft hammer, the internal components are loose and the converter should be replaced. When a catalytic converter is restricted, a significant loss of power and limited top speed will be noticed.

Various tests are available to determine if the converter is functional. One test is to measure converter inlet and outlet temperatures with a temperature probe to see if the converter lights off. A properly operating converter should show a temperature increase at the outlet compared to the inlet. Most manufacturers call for a 10 percent increase in temperature. A temperature increase greater than several hundred degrees could indicate the converter is working too hard due to excessive hydrocarbons present in the exhaust.

The engine should be checked for misfiring. A technician can use an exhaust analyzer to perform a cranking CO_2 test, an oxygen storage test, or take measurements before and after the converter to calculate converter efficiency, also called an intrusive test.

Intrusive converter testing should produce efficiency results above 80 percent. The oxygen storage test should show less than a 1.2 to 1.7 percent increase in oxygen readings during snap-throttle acceleration from 2,000 rpm if the converter is working properly and storing oxygen. A cranking CO_2 test determines if a preheated converter can convert hydrocarbons to carbon dioxide while the engine is cranked with the ignition system disabled. The fuel system must be supplying fuel because that is the source of the hydrocarbons. Tailpipe CO_2 reading should be above 12 percent and hydrocarbons should stay below 500 parts per million (ppm).

Task D.4 Evaporative Emissions Controls (3 Questions)

Task D.4.1 Test and diagnose emissions or drivability problems caused by the evaporative-emissions control system.

The evaporative-emissions control system captures and stores vapors from the fuel system in a charcoal canister to be burned once the engine is started. These vapors are purged from the canister by engine vacuum after the engine is started and run off idle.

If the engine purges vapors from the charcoal canister during idle, a rich condition and possible rough idle could result. Fuel-saturated charcoal canisters may cause excessively rich air-fuel ratios during acceleration that may cause state emission test failures. Leaks in the evaporative-emission system can cause customer complaints of gasoline odors in or around the vehicle.

Some later model OBD-II vehicles have an enhanced EVAP system that can monitor purge flow rate as well as determine if the system has fuel vapor leaks. Problems with system purging or leaks will set diagnostic trouble codes on these systems.

Task D.4.2 Interpret evaporative emission-related diagnostic trouble codes (DTCs); determine needed repairs.

Once diagnostic trouble codes are retrieved, the technician needs to consult the service manual for proper diagnostic procedures and system operation. Trouble codes will identify whether the problem is with canister fuel vapor purging or system leaks.

Due to the fact that the powertrain control module (PCM) can identify very small leaks and store a trouble code, special test equipment may be needed to diagnose and locate the problem.

Task D.4.3 Inspect, test, and replace canister, lines/hoses, mechanical and electrical components of the evaporative-emissions control systems.

A careful visual inspection of the evaporative system should be performed if a customer complaint of gasoline odors is received. A gas analyzer or smoke machine will help identify any leaks. Often hoses will be damaged or left disconnected after other repairs are done. EVAP system purge and vent solenoids need to be checked for proper resistance and electrical operation, as well as proper mechanical operation.

Most purge solenoids are normally closed and block engine vacuum to the canister when off. Energizing the purge solenoid will allow vacuum through to purge the canister. Vent solenoids are normally open and allow air to pass through the solenoid until the solenoid is energized. The vent solenoid is used to seal the system so the PCM can test for system leaks. Charcoal canisters may be equipped with filters that should be replaced at the manufacturer's specified interval.

The fuel tank cap should be carefully inspected for proper application and sealing if a system leak code is set. This is one of the most common problems setting EVAP system leak codes. Special testers are available to test the pressure and vacuum valves in the gas cap. Fuel tanks may also be equipped with a rollover valve to prevent fuel from escaping through the evaporative system if an accident caused the vehicle to turn upside down or rollover.

Some evaporative systems may use a tank pressure control valve (TPCV) that controls the flow of vapors to the charcoal canister. If fuel vapor pressure in the fuel tank is below 1.5 in. Hg, the valve will be closed and fuel vapors will be stored in the tank. When vapor pressure exceeds the set point of the valve, the vapors are vented to the charcoal canister. The TPCV also provides vacuum relief to protect against vacuum build-up in the fuel tank.

E. Computerized Engine Controls Diagnosis and Repair (including OBD II) (16 Questions)

Task E.1 Retrieve and record diagnostic trouble codes (DTC) and freeze frame data if applicable.

Retrieving diagnostic trouble codes varies greatly among the many different manufacturer's vehicles. Manual code retrieval on pre-OBD-II vehicles include some of the following procedures: installing a jumper wire across the proper terminals of a diagnostic connector and counting lamp flashes; cycling the ignition switch on and off three times in a five-second period to signal the PCM to enter diagnostic mode and counting lamp flashes; or turning a switch on the PCM to enter diagnostics and counting the blinking of LEDs in the PCM. Many pre-OBD-II vehicles also support scan tool code retrieval.

OBD-II vehicles require the use of a scan tool to retrieve and clear diagnostic trouble codes. The scan tool will also allow viewing freeze frame data that is stored in the PCM when a diagnostic trouble code is set.

Task E.2 Diagnose the causes of emissions or drivability problems resulting from failure of computerized engine controls with diagnostic trouble codes (DTC)

Once a fault has been detected by the powertrain control module (PCM), it stores a DTC in memory, and if the fault affects exhaust emissions, it will light the malfunction indicator lamp (MIL). The technician then retrieves the stored DTC and accesses a diagnostic flow chart. The diagnostic flow chart leads the technician through a series of steps to determine the actual problem.

When diagnosing a fault, it is useful to be aware of the specific set of circumstances that cause the control module to set a fault. This information is found in the workshop manual. It will allow the technician to further refine the diagnostics necessary to solve the problem.

An important step in the process of diagnosing computer trouble codes is to determine if the code is a history (memory) code or if the code is current, which means the fault is present at the time the trouble code was retrieved. Following a trouble code flow chart when a code is a history code can cause misdiagnosis and replacement of unnecessary parts. Most late-model PCMs will report codes as either current or history.

On early model systems that do not differentiate between current and history codes, the code should be cleared and the vehicle driven to see if the code resets. On OBD-II systems, the codes should not be cleared until the vehicle is repaired to prevent the PCM from erasing stored freeze frame data.

Task E.3 Diagnose the causes of emissions or drivability problems resulting from failure of computerized engine controls with no diagnostic trouble codes (DTC)

Vehicles with computerized engine control systems may exhibit many drivability or emissions problems without setting a diagnostic trouble code. A hesitation on acceleration can be experienced from a faulty throttle position sensor (TPS). A bad spot in the sensor circuit board could be causing the signal voltage to momentarily drop. The computer interprets this as a decrease in throttle position, when actually the vehicle is still accelerating. The computer may never set a diagnostic trouble code based on this type of fault because the voltage never varies above or below the voltage levels needed to set a TPS code.

A leak in the vacuum hose to a manifold absolute pressure (MAP) sensor may cause a higher-than-normal reading from the MAP sensor. The computer interprets the reading as higher engine load and increases fuel delivery. Higher exhaust emissions will result but a DTC may not be stored because the MAP sensor voltage output is still within the normal operational range. These conditions require careful pinpoint testing by the technician to identify the root cause of the problem. An understanding of normal system readings and specifications is needed so the technician can diagnose problems that do not set trouble codes in the computer. Specialty tools such as power graphing multimeters, labscopes with record modes, or graphing scan tools make the job of finding intermittent circuit or sensor problems much easier.

Task E.4 Use a scan tool, digital multimeter (DMM), or digital storage oscilloscope (DSO) to inspect or test computerized engine control system sensors, actuators, circuits, and powertrain control module (PCM); determine needed repairs.

In order to control engine operation, the PCM must have a certain set of sensor inputs on which to make decisions. Once these inputs are received, the PCM processes the signals and decides which course of action to take. The PCM then outputs signals to a series of actuators that in turn provide the engine with the things it needs to operate efficiently. In order to troubleshoot a system effectively, the technician must confirm that the sensor is transmitting the appropriate signal to the PCM and the PCM is transmitting the proper signal to the actuator.

To properly diagnose today's vehicles, a technician must be familiar with and be able to operate sophisticated diagnostic equipment such as various scan tools and DSOs. The scan tool allows a convenient means of accessing computer sensor data as well as the output commands or status of the engine control system. Many scan tools today allow bi-directional testing where the technician can take direct control of items like idle speed, or perform testing such as engine power balance, by disabling individual fuel injectors. DMM and DSO allow detailed circuit and component testing. Computer and sensor power feeds, grounds, and signal wires can be tested with a DMM. Sensor and actuator waveform analysis is performed with a DSO. Items such as fuel injectors and oxygen sensors are tested most effectively through waveform analysis. Consult equipment and manufacturer's test procedures to utilize these test instruments to their fullest capabilities.

Task E.5 Measure and interpret voltage, voltage drop, amperage, and resistance using digital multimeter (DMM) readings.

Using a DMM, the technician can evaluate circuit integrity by testing available voltage, voltage drop, amperage, and resistance readings. Available voltage tests confirm if a component is receiving the proper amount of voltage. A voltage drop test is used to locate unwanted circuit resistance when the circuit is energized. Amperage tests can locate shorted or high-resistance actuators. Resistance measurements can be used to compare components to specifications or for testing wiring harness continuity. Connecting a DMM to measure available voltage is done by making a parallel connection across the circuit. The red lead is connected at the desired test point and the black lead is connected to the negative battery terminal or engine block.

When a voltage drop test is performed, the DMM leads are connected across a component or section of a circuit on the same side of the circuit, either the power or ground side.

The reading on the voltmeter indicates the amount of voltage used by the component or section of the circuit between the meter leads. Current must be flowing in the circuit for a voltage drop to occur. A high voltage drop across a conductor or connector indicates excessive resistance. A low voltage drop across a load means resistance is present elsewhere in the circuit. Voltage drop testing across the power and ground sides of a circuit will identify the location of the unwanted resistance.

Inserting the meter in series with the circuit performs amperage testing with a DMM. All DMMs are rated for the amount of current they can measure directly. Take care are to prevent connecting the meter into a circuit with higher current draw than the meter can measure. Meters are protected with an internal fuse that will open and require replacement if the current rating of the meter is exceeded. Current probes are available that will allow a DMM to make high current measurements. Resistance measurements can be made on components that are removed from the circuit. Whenever resistance tests are performed on circuit wiring, remove battery power from the circuit under test. All ohmmeters are self-powered and must not be used on live circuits.

Task E.6 Test, remove, inspect, clean, service, and repair or replace power and ground distribution circuits and connections.

The power distribution circuit is the power and ground circuits from the battery, through the ignition switch and fuses, to the individual circuits on the vehicle. Connections must be free of corrosion, as it adds unwanted resistance to current flow.

Task E.7 Practice recommended precautions when handling static-sensitive devices and/or replacing the powertrain control module (PCM).

There are many static-sensitive components used on vehicles including powertrain, body, transmission, and ABS control computers as well as electronic instrument clusters. Static-sensitive components are shipped in an antistatic envelope. This envelope should not be opened until you are ready to install the component. Locate a good ground on the vehicle and connect a grounding strap from you to the vehicle ground before installing the component. Do not handle the component unnecessarily, and do not move around on the vehicle seat when installing the component.

Task E.8 Diagnose drivability and emissions problems resulting from failures of interrelated systems (such as cruise control, security alarms/theft deterrent, torque controls, traction controls, torque management, A/C, non-OEM installed accessories).

Today's vehicles have multiple computers with multiple functions. The computers have the ability to communicate with each other. One computer receives some sensor inputs, and the signal is forwarded to other computers. If this signal is not received, this can be interpreted as an incorrect input and may cause output problems from the processor. Theft deterrent systems may cause stalling or no-start conditions that can be difficult to trace or lead a technician toward performing unnecessary testing and parts replacement. This may be especially true with aftermarket alarm systems. Often when control modules that communicate with each other like the PCM and theft deterrent module are replaced, re-learn procedures may need to be performed before normal operation will occur.

Other drivability concerns can occur from problems in the cruise control or traction control systems such as surging or loss of power. Electrical interference may result from certain aftermarket-installed accessories such as stereo amplifiers. Improper installation or wiring damage is also a concern when non-OEM components are installed.

A technician must identify all systems that are used on a vehicle and what possible interaction they may have on the PCM. The vehicle service manual and service bulletins should be consulted when problems are suspected in interrelated systems.

Task E.9 Diagnose the causes of emissions or drivability problems resulting from computerized spark timing controls; determine needed repairs.

The ignition module receives an input from a Hall-effect pickup or a variable reluctance sensor; this signal is used to fire the coil(s) on startup. The ignition module sends this signal to the powertrain control module (PCM), and the PCM interprets it as a revolutions per minute (rpm) input. This signal between the ignition module and PCM is a digital signal. The PCM then sends a varying digital signal back to the ignition module. The module uses this signal as a computed timing signal and fires the coil(s) based on this information.

Problems with spark timing controls will often set diagnostic trouble codes. If the computer detects a spark timing control circuit problem, the engine will operate at base ignition timing and a lack of power may result. Improper sensor inputs may cause changes in spark timing control and lead to increased emissions. For example, a coolant sensor that is out of calibration and sends a colder-than-actual temperature reading may cause the PCM to increase spark timing. This, in turn, will increase fuel delivery and spark advance resulting in excessive fuel consumption and emissions.

Task E.10 Verify the repair, and clear diagnostic trouble codes (DTCs).

Prior to the introduction of OB -II, each manufacturer had its own method for erasing DTCs from the memory of a PCM. These procedures should always be followed. Normally, verification of the repair is done by operating the engine and the related system and checking to see if the operation triggered the DTC. If it did not, the problem was probably solved.

On OBD II-equipped vehicles, the fail records and the freeze frame data for the DTC that was diagnosed should be reviewed and recorded. Then use a scan tool's clear DTCs or clear info functions to erase the DTCs from the PCM's memory. Operate the vehicle within the conditions noted in the fail records and/or the freeze frame data. Then monitor the status information for the specific DTC until the diagnostic test associated with that DTC runs.

F. Engines Electrical Systems Diagnosis and Repair (4 Questions)

Task F.1 Battery (1 Question)

Task F.1.1 Test and diagnose emissions or drivability problems caused by battery condition, connections, or excessive key-off battery drain; determine needed repairs.

Battery voltage is not only needed to start the engine, but is also very important in stabilizing the voltage during engine operation. Low battery voltage or state of charge, as well as poor battery cable connections, will cause slow engine cranking and hard or no-start complaints. Once the engine is running, the vehicle's charging system supplies the voltage needed and restores the charge of the battery. As the charge voltage is supplied to the system, it will fluctuate depending on vehicle electrical loads and the sensed need of the battery. Once the battery has the same voltage as the output of the charging system, the system voltage and charging output is leveled. Many systems on today's vehicles are monitored by electronic components that send information to a control module or computer. This information is delivered as changes in voltage. Therefore, precise voltage control is important to effective engine control management. When there is a large fluctuation in system voltage or voltage spikes, the computer may reset base sensor input levels stored in memory and drivability problems may result.

Most computers have a few milliamperes of current draw when they are not in operation. This current draw is called parasitic load. Since many vehicles today have several computers, this current draw may discharge a battery if the vehicle is not driven for several weeks. Most manufacturers allow between 50 to 100 milliamps parasitic current draw. A value greater than 100 milliamps is considered excessive and further diagnostics should be performed to determine what is causing the current drain.

Connecting a DMM set to measure amperage in series between the negative battery cable and battery post will allow the technician to measure this load. All accessories must be off and the key removed from the ignition. Some vehicles may require a long wait period of up to 1 hour before all computers will power down. Take a reading and compare it to manufacturer's specifications.

Task F.2 Starting System (1 Question)

Task F.2.1 Perform starter current draw test; determine needed action.

Starter current draw testing is only performed on batteries with a specific gravity of 1.190 or greater. Several different testers can be used. If an analog tester is used, always check the mechanical zero on each meter and adjust as necessary. Be sure that all electrical loads are off and the doors are closed, as additional loads will cause additional draw. The ignition is disabled and the engine is cranked while observing the ammeter and voltmeter readings. High current draw and low cranking speed usually indicate a defective starter. High current draw may also be caused by internal engine problems. A low cranking speed and low current draw with high cranking voltage usually indicate excessive resistance in the starter circuit, such as in the cables and connections.

Task F.2.2 Perform starter circuit voltage drop tests; determine needed action.

The resistance in an electrical wire may be checked by measuring the voltage drop across the wire with normal current flow in the wire. To measure the voltage drop across the positive battery cable, connect the positive voltmeter lead to the positive cable at the battery, and connect the negative voltmeter lead to the other end of the positive battery cable at the starter solenoid. Disable the ignition system. Crank the engine. The voltage drop indicated on the meter should not exceed 0.5V. If the voltage reading is above this figure, the cable has excessive resistance. If the cable ends are clean and tight, replace the cable. Connect the positive voltmeter lead to the positive battery cable on the starter solenoid, and connect the negative voltmeter lead to the starting motor terminal on the other side of the solenoid. Leave the voltmeter on the lowest scale and crank the engine.

If the voltage drop exceeds 0.3V, the solenoid disc and terminals have excessive resistance.

Connect the positive voltmeter lead to the starter motor housing and the negative lead to the negative battery post and crank the engine. If the reading is greater than 0.2 volts and the cable connections are clean and tight, replace the negative battery cable.

Task F.2.3 Inspect, test, and repair or replace components and wires in the starter control circuit.

When testing the starter control circuit, connect the positive voltmeter lead to the positive battery cable at the battery, and connect the negative voltmeter lead to the solenoid winding terminal on the solenoid. Leave the ignition system disabled and place the voltmeter selector on the lowest scale. If the voltage drop across the control circuit exceeds 1.5V while cranking the engine, individual voltage drop tests on control circuit components are necessary to locate the high resistance problem.

Task F.3 Charging System (2 Questions)

Task F.3.1 Test and diagnose engine performance problems resulting from an undercharge, overcharge, or a no-charge condition; determine needed action.

The charging system is responsible for maintaining stable electrical system voltage. Undercharging as well as overcharging may cause engine performance problems including hard starting or battery gassing and corrosion concerns. A defective diode in the alternator may allow enough AC voltage leakage into the electrical system to disrupt normal computer operation. Test the alternator with a DMM or labscope for excessive AC voltage output and replace it if found out of specifications.

If alternator output is zero, the alternator field circuit may be open. The most likely place for an open circuit in the alternator is at the slip rings and brushes. If alternator output is normal but no charging current is measured at the battery, the fuse link between the alternator output terminal and positive battery cable is probably open. If the alternator output is less than specified, always be sure that the belt and belt tension are satisfactory. If belt condition and tension is satisfactory and the alternator output is less than specified, the alternator is defective.

On some alternators, there is a method of "full fielding" the unit. This technique will bypass the voltage regulator circuit and full alternator output will be obtained. In this case, if the alternator has full output, then the regulator or its circuit has failed.

Task F.3.2 Inspect, adjust, and replace alternator (generator) drive belts, pulleys, tensioners, and fans.

A loose belt causes low alternator output and a discharged battery. A loose, dry, or worn belt may cause squealing and chirping noises during acceleration and cornering. Belt tension may be checked by measuring the belt deflection. Press on the belt with the engine stopped to measure the belt deflection. 1⁄2 in (12.7 mm) per foot (30.5 cm) of free span is usually acceptable.

Serpentine drive belt systems will most often use automatic belt tensioners. Automatic tensioners are usually spring loaded, but should be checked to be sure they are applying adequate pressure to the belt to prevent slipping or squealing from the belt. Pulley alignment is critical on serpentine belt drives. Any misalignment will cause noise from the belt and may allow the belt to slip off the pulleys.

Whenever a customer with a serpentine drive belt engine consistently complains of belt noise, the technician should check to see if any previous service has been performed on the engine that may have required accessories to be removed. If a pulley is reinstalled backward or any spacers are left off of accessory brackets, pulley misalignment can occur and cause constant belt noise. Proper routing of the serpentine belt must be observed when replacing a drive belt. Routing the belt incorrectly can cause some accessories to spin backward or not allow proper tension to be applied to the belt by the belt tensioner.

Task F.3.3 Inspect, test, and repair or replace charging circuit connectors and wires.

Check wires for burned or melted conditions. Check connector ring terminals for loose retaining nuts, which cause high resistance or intermittent open circuits. An open circuit may be caused by a terminal that is backed-out of the connector. Terminals that are bent or damaged may cause shorts or open circuits. An open circuit occurs when the terminal is crimped over the insulation instead of the wire core. A greenish white corrosion on terminals results in high resistance or an open circuit.

5 Sample Test for Practice

Sample Test

Please note the letter and number in parentheses following each question. They match the task in Section 4 that discusses the relevant subject matter. You may want to refer to the overview using the cross-referencing key to help with questions posing problems for you.

1. While testing a turbocharger, the maximum boost pressure observed is 4 psi (27.6 kPa), while the specified pressure is 9 psi (62 kPa). Technician A says the engine compression may be low. Technician B says the wastegate may be sticking open. Who is right?
   A. Technician A only
   B. B only
   C. Both A and B
   D. Neither A nor B (A.15)

2. All of the following are measured by a four-gas analyzer **EXCEPT:**
   A. hydrocarbons (HC).
   B. carbon monoxide (CO).
   C. oxides of nitrogen (NOx).
   D. oxygen (O_2).
   (A.10)

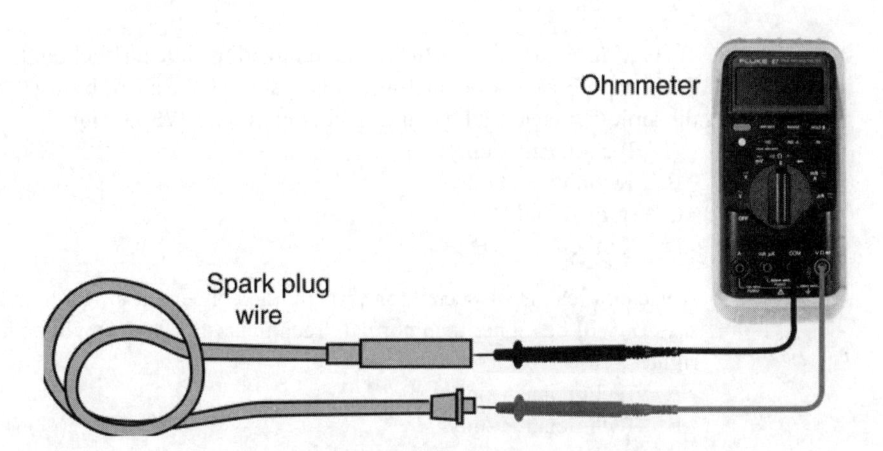

3. While testing a spark plug wire as shown in the figure, Technician A says the manufacturer has a specified amount of resistance for each foot of plug wire. Technician B says plug wire resistance readings should indicate no (zero) resistance. Who is right?
   A. A only
   B. Technician B only
   C. Both A and B
   D. Neither A nor B (B.5)

4. When diagnosing fuel-injection system problems, a technical service bulletin search is performed for all of the following reasons **EXCEPT:**
   A. to save diagnostic time.
   B. to locate service manual updates or specification changes.
   C. mid-year production changes.
   D. year, make, and model identification. (A.2)

5. Technician A says that on carburetor-equipped engines idle mixture is set before idle speed and ignition timing. Technician B says an inspection of associated vacuum lines and electrical connectors should be done before idle speed and timing adjustments are attempted. Who is right?
   A. Technician A only
   B. B only
   C. Both A and B
   D. Neither A nor B (B.7 and C.11)

6. Which of the following conditions is LEAST-Likely to occur from replacing a battery in a late-model vehicle?
   A. Radio will not play
   B. Engine stalling or erratic idle
   C. Air conditioning inoperative
   D. Transmission shift quality concerns (A.2 and F.1.1)

7. A test lamp is connected between the negative side of the coil and ground to diagnose a no-start condition. Technician A says a flickering test lamp could be caused by a defective ignition module. Technician B says a flickering test lamp could be caused by a defective pickup coil. Who is right?
   A. Technician A only
   B. Technician B only
   C. Both A and B
   D. Neither A nor B (B.1 and B.3)

8. Technician A says there is no provision to adjust idle air-fuel mixture on a fuel-injected engine. Technician B says some fuel-injection systems have an air by-pass screw or adjustment screw on the airflow meter to adjust idle air-fuel mixtures. Who is right?
   A. Technician A only
   B. Technician B only
   C. Both A and B
   D. Neither A nor B (A.2)

9. A vacuum leak is suspected for a rough idle concern. Using a four-gas analyzer, Technician A says O_2 will be higher than normal. Technician B says CO will be higher than normal. Who is right?
   A. Technician A only
   B. Technician B only
   C. Both A and B
   D. Neither A nor B (A.10 and C.1)

10. If the vacuum drops slowly to a low reading when a vacuum gauge is connected to the intake manifold and the engine is accelerated and held at a steady speed, it indicates:
    A. sticking valves.
    B. over-advanced ignition timing.
    C. a restricted exhaust.
    D. a rich fuel mixture. (A.5 and C.14)

11. Technician A says a thorough ignition coil test includes both primary and secondary winding resistance tests. Technician B says maximum coil output testing can be performed with an oscilloscope. Who is right?
    A. Technician A only
    B. Technician B only
    C. Both A and B
    D. Neither A nor B (B.6)

12. A proper cooling system inspection involves all of the following **EXCEPT** a:
    A. pressure test.
    B. cooling fan inspection.
    C. evaporator core inspection.
    D. thermostat inspection. (A.13)

13. While performing engine noise diagnostics, Technician A says a stethoscope is a good tool for noise location. Technician B says you may have to duplicate the specific operating condition. Who is right?
    A. Technician A only
    B. Technician B only
    C. Both A and B
    D. Neither A nor B (A.3)

14. When replacing a PROM, Technician A says that you should never ground yourself to the vehicle. Technician B says that grounding yourself to the vehicle will erase the PROM. Who is right?
    A. Technician A only
    B. Technician B only
    C. Both A and B
    D. Neither A nor B (E.7)

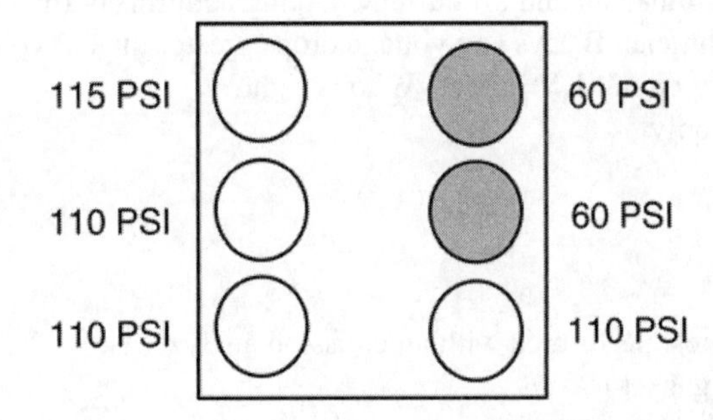

15. Technician A says that during a cylinder compression test, low readings on adjacent cylinders, as shown in the figure, may be caused by a blown cylinder head gasket. Technician B says a low reading on single cylinder is probably a piston ring or valve problem. Who is right?
    A. Technician A only
    B. Technician B only
    C. Both A and B
    D. Neither A nor B (A.7)

16. The first step a technician should take when performing a diagnostic procedure is:
    A. scan test for trouble codes.
    B. verify the customer concern.
    C. test the battery state of charge.
    D. consult manufacturer service bulletins. (A.1)

17. Technician A says the vacuum lines at the distributor are usually disconnected and plugged to check base ignition timing. Technician B says the distributor rotor must point at the specified cylinder's distributor cap terminal when installing the distributor. Who is right?
    A. Technician A only
    B. Technician B only
    C. Both A and B
    D. Neither A nor B (B.4 and B.7)

18. When installing and timing the distributor, Technician A says the engine must be timed referencing TDC on the specified cylinder's compression stroke. Technician B says if the engine is timed on the compression stroke, the distributor will be 180 degrees off. Who is right?
    A. A only
    B. Technician B only
    C. Both A and B
    D. Neither A nor B (B.7)

19. Technical service bulletins should be used for all of the following **EXCEPT:**
    A. to locate service manual updates.
    B. to find updated parts or service procedures.
    C. to find updated computer PROM or calibration information.
    D. to look up engine torque specifications. (A.2)

20. Technician A says that a 12-volt test light connected between the negative side of an ignition coil and ground that blinks on and off during cranking confirms normal operation of the primary ignition circuit. Technician B says any voltage drops greater than .2 volts in the primary circuit can reduce secondary circuit KV output. Who is right?
    A. Technician A only
    B. B only
    C. Both A and B
    D. Neither A nor B (B.3)

21. The LEAST-Likely test performed with an emission analyzer is:
    A. cylinder head gasket leak.
    B. O_2 sensor waveform analysis.
    C. cylinder misfire.
    D. inspection and maintenance program exhaust analysis. (A.10)

22. An engine has a lack of power and excessive fuel consumption. Technician A says a broken timing belt cannot be the cause. Technician B says the timing belt may have jumped a tooth. Who is right?
    A. Technician A only
    B. Technician B only
    C. Both A and B
    D. Neither A nor B (A.12)

23. While discussing exhaust color diagnosis, Technician A says black smoke in the exhaust indicates a lean air-fuel mixture. Technician B says white smoke in the exhaust indicates oil leakage into the combustion chamber. Who is right?
    A. Technician A only
    B. Technician B only
    C. Both A and B
    D. Neither A nor B (A.4)

24. Two technicians are discussing fuel-injector pressure balance testing. Technician A says an injector with a higher pressure drop indicates a rich running injector. Technician B says an injector with a higher pressure drop could be leaking. Who is right?
    A. Technician A only
    B. Technician B only
    C. Both A and B
    D. Neither A nor B (C.8)

25. While testing the cooling system, Technician A says to repeat the pressure test after repairs are made to ensure that all leaks are found. Technician B says a pressure test should include testing the radiator cap. Who is right?
    A. Technician A only
    B. Technician B only
    C. Both A and B
    D. Neither A nor B (A.13)

26. Valve adjustment is being discussed. Technician A says valve adjustment should always be performed on a cold engine. Technician B says the piston should be placed at TDC of the compression stroke. Who is right?
    A. Technician A only
    B. Technician B only
    C. Both A and B
    D. Neither A nor B (A.11)

27. A vehicle with a SFI V6 engine and OBD-II emissions controls has set a DTC PO171 (System Too Lean, Bank 1). No other drivability concerns are present. The freeze frame data shows the code was set under warm idle conditions. Technician A says the problem could be an intake manifold vacuum leak. Technician B says the problems could be a weak fuel pump. Who is right?
    A. A only
    B. Technician B only
    C. Both A and B
    D. Neither A nor B (C.2, C.4, C.10 and E.1)

28. All are inputs for timing control, **EXCEPT:**
    A. knock sensor.
    B. engine speed (rpm).
    C. power steering load.
    D. engine load. (A.2 and B.7)

29. To check a coil's available voltage output, the technician should:
    A. disconnect the fuel pump power lead.
    B. disconnect the plug wire at the plug and ground it.
    C. disconnect the coil wire and ground it.
    D. conduct the test using a suitable spark tester that requires 25 kV. (B.1 and B.6)

30. A car with an AIR system backfires on deceleration. The technician should check:
    A. the air manifold for restrictions.
    B. operation of the diverter or gulp valve.
    C. operation of the exhaust manifold check valve.
    D. output pressure of the air pump. (D.3.1 and D.3.3)

31. A cylinder power balance test can indicate all of the following problems **EXCEPT:**
    A. a bad spark plug.
    B. late ignition timing.
    C. an open ignition wire.
    D. burned valves. (A.6)

32. In discussing a vehicle that has an engine miss under acceleration, and sometimes at cruise speed, but idles smooth, Technician A says a coil with weak available voltage could cause it. Technician B says there may be an engine misfire DTC stored to aid in diagnosis. Who is right?
    A. Technician A only
    B. Technician B only
    C. Both A and B
    D. Neither A nor B (B.1, B.2 and B.6)

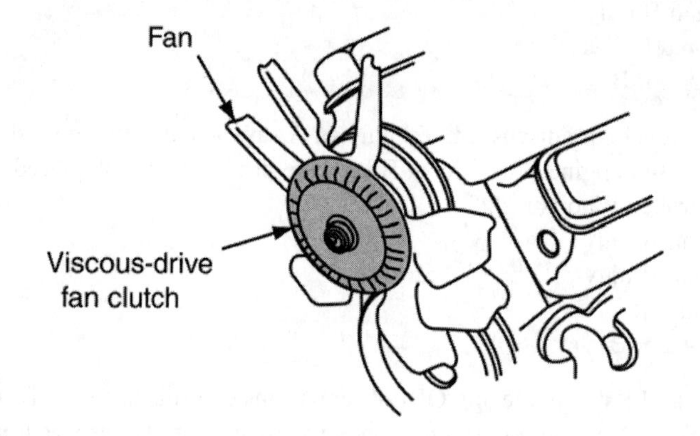

33. When testing a viscous-drive fan clutch, as shown in the figure, with the engine off, rotate the cooling fan by hand. It should have:
    A. more resistance hot.
    B. more resistance cold.
    C. no rotation movement.
    D. no resistance. (A.14)

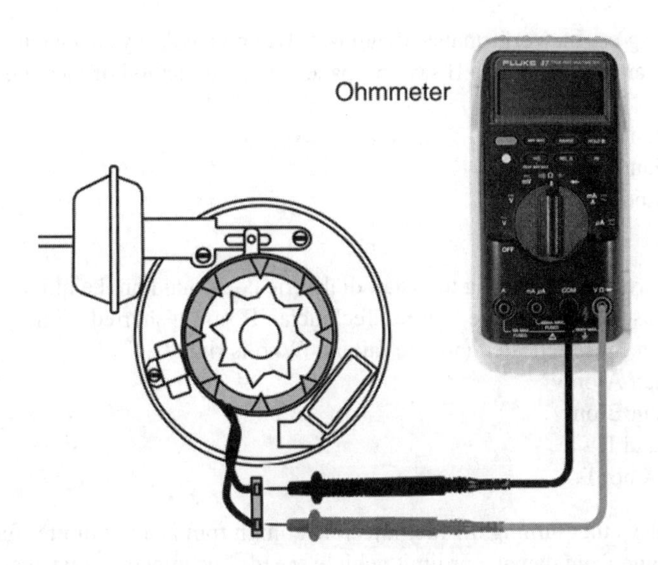

34. When testing a pickup coil with an ohmmeter, as shown in the figure, Technician A says when the pickup coil leads are moved, an erratic ohmmeter reading is normal. Technician B says that an infinite ohmmeter reading between the pickup coil terminals is an acceptable reading. Who is right?
    A. Technician A only
    B. Technician B only
    C. Both A and B
    D. Neither A nor B

(B.8)

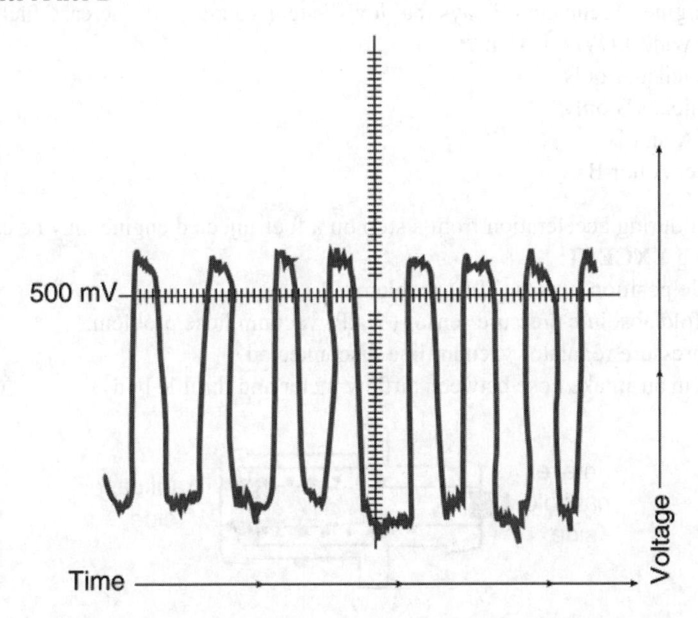

35. Based on the O_2 sensor waveform shown, all the following are true **EXCEPT:**
    A. this represents a lean biased condition.
    B. a diagnostic trouble code may be recorded in the PCM.
    C. this represents a rich biased condition.
    D. the O_2 sensor is functioning.

(E.3 and E.4)

36. While discussing engine performance diagnosis, Technician A says a vacuum leak decreases engine performance. Technician B says propane is a good method of locating vacuum leaks. Who is right?
   A. Technician A only
   B. Technician B only
   C. Both A and B
   D. Neither A nor B (C.10)

37. Technician A says that forgetting to install dielectric compound on the mounting surface of the module can cause repeat module failure. Technician B says a shorted primary winding in the ignition coil can cause repeated module failure. Who is right?
   A. Technician A only
   B. Technician B only
   C. Both A and B
   D. Neither A nor B (B.9)

38. Technician A says that turning off fuel injectors at high rpm is a rev limiter function used to protect the engine from damage or limit vehicle speed. Technician B says that turning off fuel injectors while the engine is running is not done on any fuel-injection systems. Who is right?
   A. A only
   B. Technician B only
   C. Both A and B
   D. Neither A nor B (C.1)

39. Two technicians are discussing fuel-injection system operating strategies. Technician A says that cranking the engine with the throttle fully depressed will force a lean mixture to clear a flooded engine. Technician B says that low-system voltage will increase fuel injector on time (pulse width). Who is right?
   A. Technician A only
   B. Technician B only
   C. Both A and B
   D. Neither A nor B (A.2 and C.1)

40. A hesitation during acceleration from a stop on a fuel-injected engine may be caused by all of the following **EXCEPT:**
   A. throttle position sensor (TPS) problem.
   B. manifold absolute pressure sensor (MAP) vacuum hose problem.
   C. fuel-pressure regulator vacuum line disconnected.
   D. crack in air intake hose between airflow meter and throttle body. (C.1, C.6 and E.3)

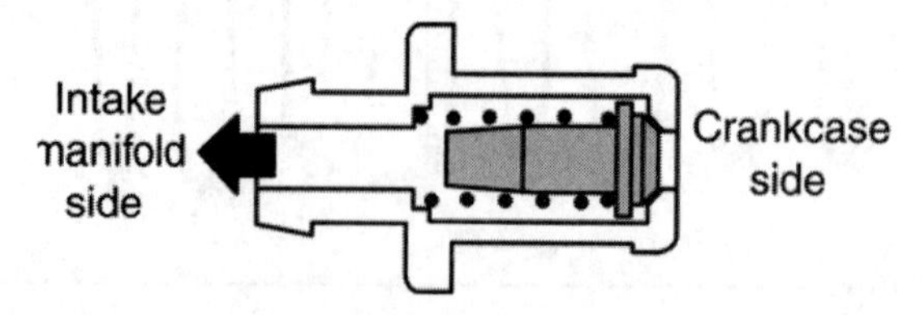

41. In the figure, the plunger is stuck in the maximum flow position. Technician A says this can cause a rough idle. Technician B says this could cause excessive oil consumption. Who is right?
   A. Technician A only
   B. Technician B only
   C. Both A and B
   D. Neither A nor B (D.1.1 and D.1.2)

42. A starter free-running test is being made on the bench with a fully charged battery. The current draw is higher than specification and the rpm is lower. Technician A says this could be caused by tight bushings. Technician B says this could be caused by worn brushes. Who is right?
    A. A only
    B. Technician B only
    C. Both A and B
    D. Neither A nor B (F.2.1)

43. The LEAST-Likely cause of a fuel tank leak is:
    A. defective tank straps.
    B. road damage.
    C. defective seams.
    D. corrosion. (C.3)

44. Nylon fuel hoses should be inspected for all **EXCEPT:**
    A. kinks.
    B. loose fittings.
    C. discoloration.
    D. scratches. (C.3)

45. While testing fuel pressure on a TBI engine, Technician A says there will always be a Shrader test port for fuel system testing. Technician B says that a high fuel pressure reading could be the result of a plugged fuel filter trapping fuel between the filter and the fuel rail. Who is right?
    A. Technician A only
    B. Technician B only
    C. Both A and B
    D. Neither A nor B (C.4)

46. Technician A says a fuel-pressure test will test fuel pump operation. Technician B says it is possible to have a good pressure reading and insufficient flow. Who is right?
    A. Technician A only
    B. Technician B only
    C. Both A and B
    D. Neither A nor B (C.6)

47. Air is escaping from the PCV valve opening during a cylinder leakage test. Technician A says a blown head gasket is the cause. Technician B says air escaping from the PCV valve opening is leaning past the ring. Who is right?
    A. A Technician only
    B. Technician B only
    C. Both A and B
    D. Neither A nor B (A.8)

48. The LEAST-Likely cause of poor fuel mileage on a vehicle with EFI is:
    A. high fuel pressure.
    B. disconnected regulator vacuum hose.
    C. partially plugged exhaust.
    D. low fuel pressure. (C.1)

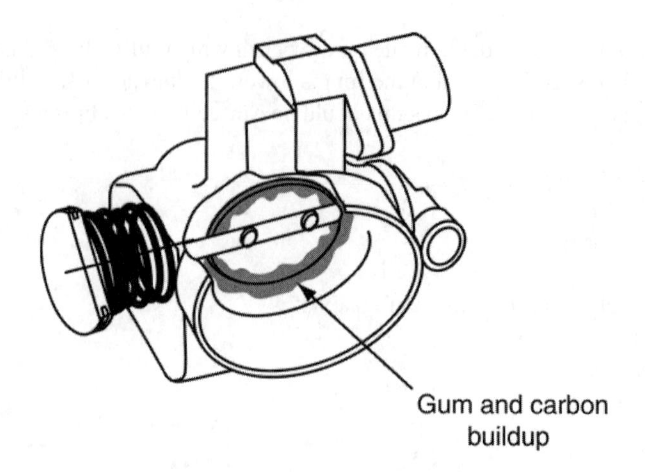

49. While discussing throttle body carbon deposits, as shown in the figure, Technician A says if approved methods do not properly clean a throttle body, the throttle body must be removed and disassembled. Technician B says a buildup of gum and carbon deposits may cause rough idle operation. Who is right?
    A. Technician A only
    B. Technician B only
    C. Both A and B
    D. Neither A nor B (C.7)

50. Technician A says performing a fuel pressure test confirms proper operation of the fuel pump. Technician B says it is possible to have a hydraulic problem with an injector, even though the electrical resistance is within specifications. Who is right?
    A. Technician A only
    B. B only
    C. Both A and B
    D. Neither A nor B (C.1, C.4 and C.8)

51. If the air-fuel mixture is ignited before the spark plug fires, this is called:
    A. preignition.
    B. over-advanced timing.
    C. dieseling.
    D. lean burn combustion. (B.1)

52. Technician A says a leaking cold start injector could cause a high CO reading at idle. Technician B says that no vacuum to the pressure regulator could cause a high CO reading at idle. Who is right?
    A. Technician A only
    B. Technician B only
    C. Both A and B
    D. Neither A nor B (A.10 and C.1)

53. The LEAST-Likely condition a manifold absolute pressure (MAP) sensor can cause is:
    A. a rich or lean air-fuel ratio.
    B. engine surging.
    C. excess fuel consumption.
    D. excessive idle speeds. (E.2 and E.3)

54. The first thing a technician should do to test a cold-start injector is to:
    A. check the spray pattern.
    B. perform injector balance testing.
    C. check the cold start injector resistance value.
    D. energize the injector and watch for a pressure drop. (C.8)

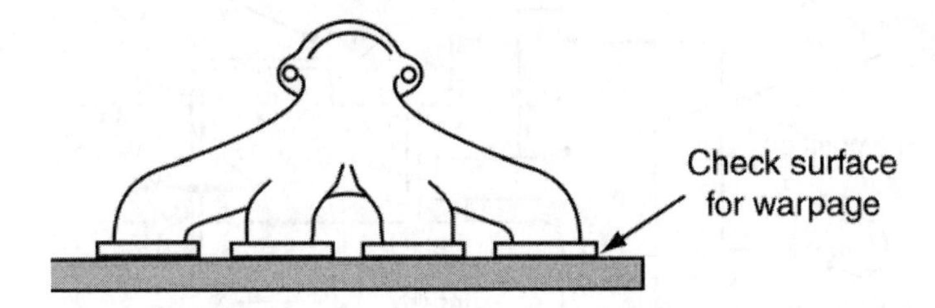

55. To check an exhaust manifold for warpage between ports, as shown in the figure, Technician A says that only a straightedge is needed. Technician B says a straightedge and a flashlight should be used. Who is right?
    A. Technician A only
    B. Technician B only
    C. Both A and B
    D. Neither A nor B (C.13)

56. Technician A says that when vacuum is applied to the exhaust gas recirculation (EGR) valve with the engine idling, the EGR valve should open and idle should become erratic. Technician B says that a diagnosis of the EGR valve should not be done with the engine idling. Who is right?
    A. A only
    B. Technician B only
    C. Both A and B
    D. Neither A nor B (D.2.1)

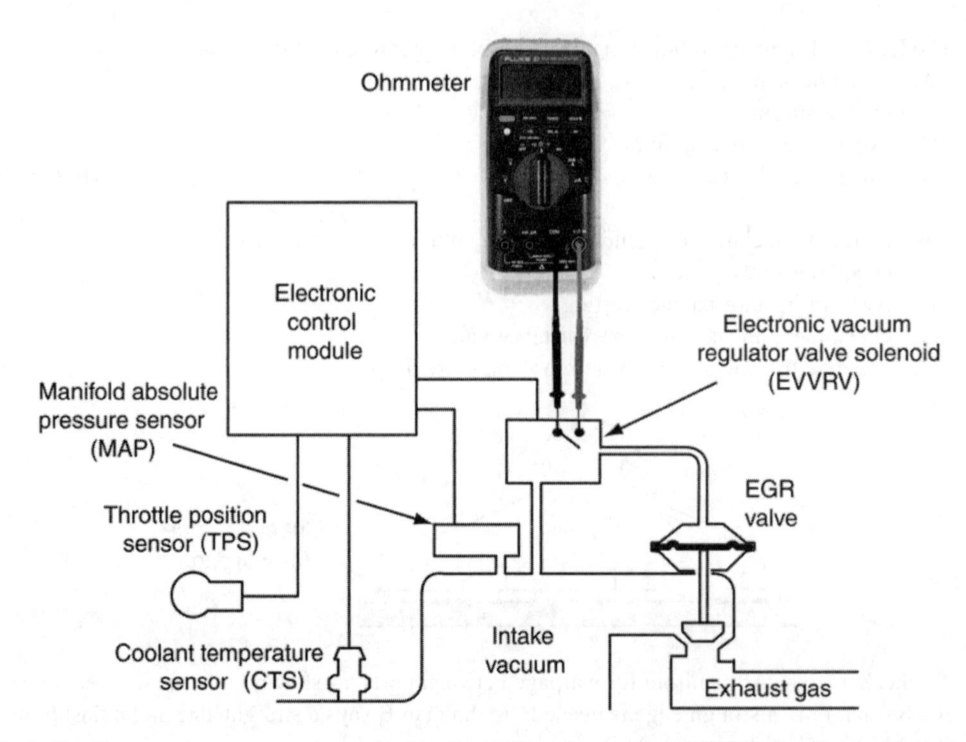

57. In the figure, when vacuum is applied to the EGR valve by the EVVRV, the diagnostic switch in the EVVRV solenoid closes. A constant infinite resistance reading between the switch contacts during normal engine operation could mean any of the following **EXCEPT:**
    A. the vacuum regulator is defective.
    B. the EGR valve is defective.
    C. normal switch operation.
    D. a broken or disconnected vacuum line to the EVVRV. (D.2.1 and D.2.3)

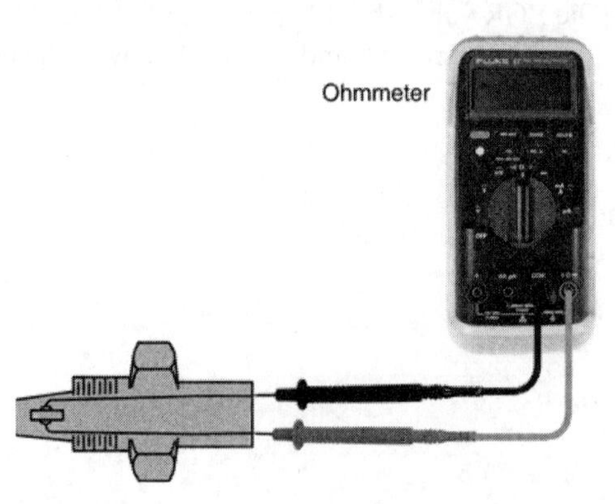

58. In the figure, the coolant sensor (a negative temperature coefficient type) resistance is being tested against specifications. Technician A says a resistance reading higher than specifications would send a signal indicating a warmer than actual engine temperature. Technician B says if the resistance is higher than specifications, the engine may exhibit hard starting when warm. Who is right?
    A. Technician A only
    B. B only
    C. Both A and B
    D. Neither A nor B (E.4 and E.5)

59. A stuck air-injection control valve that constantly sends air pump output to the exhaust manifold will MOST Likely result in:
    A. improved fuel economy.
    B. pinging on acceleration.
    C. engine overheating.
    D. a constant lean oxygen sensor signal. (D.3.1)

60. Technician A says secondary air injection is added to some catalytic converters to help oxidize hydrocarbon and carbon monoxide emissions. Technician B says secondary air is never added to a catalytic converter. Who is right?
    A. A only
    B. Technician B only
    C. Both A and B
    D. Neither A nor B (D.3.1)

61. Technician A says the evaporative-emission system assists in fuel vaporization in the intake manifold. Technician B says the evaporative-emissions system prevents fuel vapors from escaping into the atmosphere. Who is right?
    A. Technician A only
    B. B only
    C. Both A and B
    D. Neither A nor B (D.4.1)

62. Technician A says some evaporative emissions canisters have a replaceable filter. Technician B says if the filler cap is equipped with pressure and vacuum valves, they must be checked for dirt contamination and damage. Who is right?
    A. Technician A only
    B. Technician B only
    C. Both A and B
    D. Neither A nor B (D.4.3)

63. A diagnostic trouble code (DTC) for a coolant sensor may be set by any of the following conditions **EXCEPT:**
    A. an open in the voltage reference wire.
    B. operating the vehicle in extremely cold climates.
    C. a short in the voltage reference wire.
    D. an out of range sensor input. (E.1 and E.2)

64. All of the following reduce turbocharger life **EXCEPT:**
    A. inadequate cooling.
    B. lack of oil changes.
    C. lack of air cleaner maintenance.
    D. exhaust system damage. (C.15)

65. Technician A says a faulty throttle position sensor (TPS) can cause a hesitation on acceleration. Technician B says a faulty TPS will always set a diagnostic trouble code (DTC). Who is right?
    A. A only
    B. Technician B only
    C. Both A and B
    D. Neither A nor B (E.2.3, E.3 and E.4)

66. Technician A says restricted exhaust may cause reduced engine power. Technician B says restricted exhaust causes reduced maximum speed. Who is right?
    A. Technician A only
    B. Technician B only
    C. Both A and B
    D. Neither A nor B (C.14)

67. While discussing positive crankcase ventilation (PCV) system diagnosis, Technician A says with the PCV valve disconnected from the rocker cover, there should not be vacuum at the valve with the engine idling. Technician B says when the PCV valve is removed and shaken, there should not be a rattling noise. Who is right?
    A. Technician A only
    B. Technician B only
    C. Both A and B
    D. Neither A nor B (D.1.2)

68. A scan test of the computer system on a late-model fuel-injected engine reveals a bank 1 long term fuel trim value of positive 22 and a bank 2 long term fuel trim value of negative 2 with the engine idling. Technician A says these readings could be caused by a vacuum leak. Technician B says a bad oxygen sensor on bank 1 could cause these readings. Who is right?
    A. Technician A only
    B. Technician B only
    C. Both A and B
    D. Neither A nor B (E.3 and E.4)

69. Technician A says if the converter is restricted, a loss of power and limited top speed will be noticed while driving the vehicle. Technician B says if you tap on a monolithic (honeycomb) style converter with a rubber mallet and the converter rattles, it should be replaced. Who is right?
    A. Technician A only
    B. Technician B only
    C. Both A and B
    D. Neither A nor B (D.3.1 and D.3.4)

70. While discussing manifold absolute pressure (MAP) sensors, Technician A says a MAP sensor should be able to hold vacuum during a test. Technician B says some MAP sensors produce an analog voltage signal while others produce a digital square wave signal. Who is right?
    A. Technician A only
    B. Technician B only
    C. Both A and B
    D. Neither A nor B (E.3 and E.4)

71. Technician A says the oxygen sensor can be removed and tested under different temperatures. Technician B says you can use a scan tool to check for oxygen sensor codes and operation. Who is right?
    A. Technician A only
    B. B only
    C. Both A and B
    D. Neither A nor B (E.4)

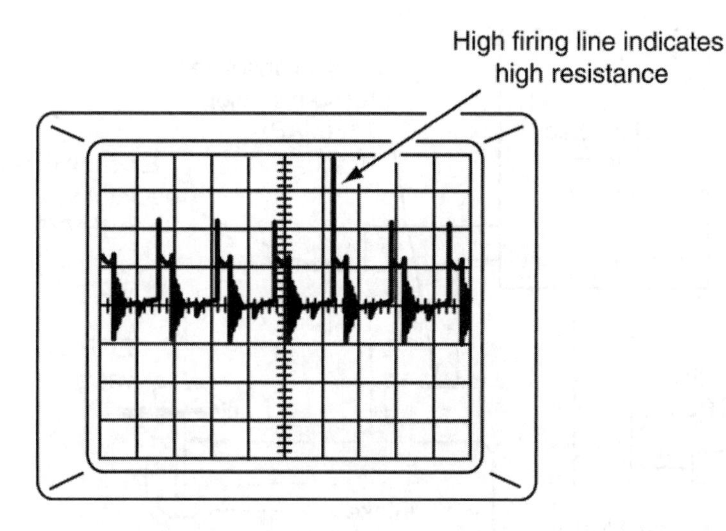

72. While monitoring secondary ignition with an oscilloscope, as shown in the figure, the LEAST-Likely cause of high resistance in the ignition secondary circuit is:
    A. damaged spark plug wires.
    B. no dielectric compound on the ignition module mounting surface.
    C. corroded spark plug wire ends.
    D. excessive spark plug air gap. (A.9, B.1 and B.5)

73. Technician A says a voltage drop test checks for excessive resistance between two test points. Technician B says more than a 0.5V voltage drop indicates excessive resistance across the battery positive cable. Who is right?
    A. Technician A only
    B. Technician B only
    C. Both A and B
    D. Neither A nor B (E.5, F.1.1 and F.2.2)

74. Technician A says that the first step of any diagnostic procedure is to check for diagnostic trouble codes (DTCs). Technician B says that the customer complaint should be verified before performing any diagnostic procedures. Who is right?
    A. Technician A only
    B. Technician B only
    C. Both A and B
    D. Neither A nor B (A.1)

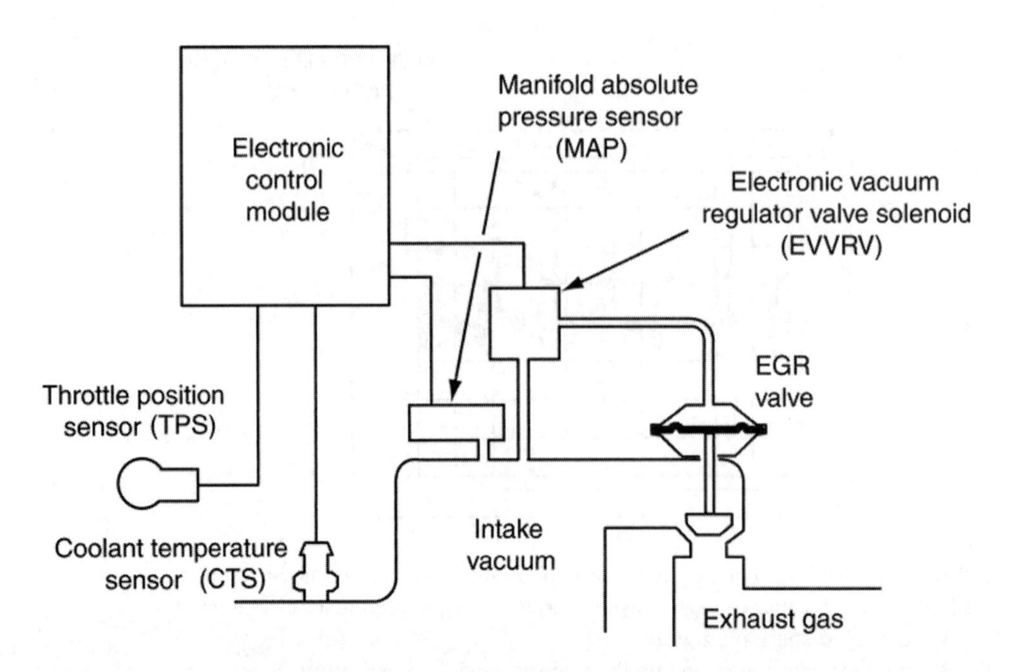

75. Technician A says that the first step in diagnosing any exhaust gas recirculation (EGR) valve concern is to check the vacuum and electrical connections. Technician B says that in many systems, as shown in the figure, the powertrain control module (PCM) uses other sensor inputs that could cause an EGR problem, and therefore diagnostic trouble codes (DTCs) should be corrected before replacing any EGR components. Who is right?
    A. Technician A only
    B. Technician B only
    C. Both A and B
    D. Neither A nor B (A.2, A.6, C.1 and C.8)

76. A spark knock from the engine on light acceleration would MOST Likely be caused by:
    A. a rich fuel mixture.
    B. a bad crankshaft sensor.
    C. high manifold vacuum.
    D. a restricted EGR. (D.2.1)

77. Technician A says that enabling criteria are the specific conditions that must be met before a monitor will run such as ambient temperature or engine load. Technician B says that pending conditions are conditions that exist that prevent a specific monitor from running, such as an oxygen sensor fault code preventing a catalyst monitor from running. Who is right?
    A. Technician A only
    B. Technician B only
    C. Both A and B
    D. Neither A nor B (A.2 and E.1)

78. Technician A says that worn valvetrain components usually produce an identifiable noise. Technician B says an engine noise diagnosis should be performed before doing engine repair work. Who is right?
    A. Technician A only
    B. Technician B only
    C. Both A and B
    D. Neither A nor B (A.3)

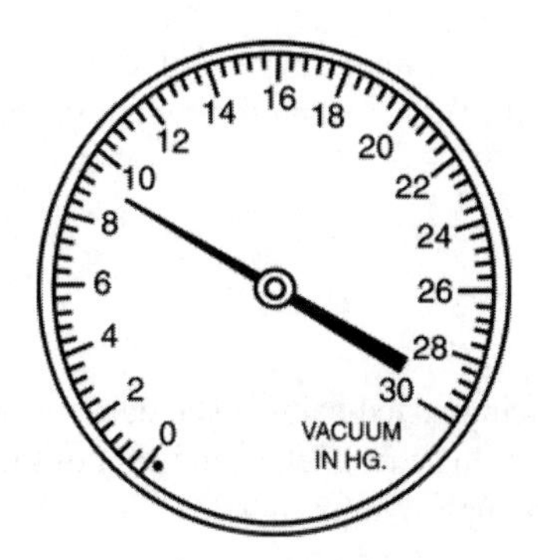

79. The vacuum gauge in the figure indicates low vacuum. Technician A says late ignition timing will cause a low vacuum reading. Technician B says to connect the gauge to a ported vacuum port. Who is right?
    A. Technician A only
    B. Technician B only
    C. Both A and B
    D. Neither A nor B (A.5)

80. The LEAST-Likely cause of low cylinder compression is:
    A. worn valves.
    B. worn rings.
    C. blown head gasket.
    D. worn valve guides. (A.7)

81. Technician A says incorrect valve timing can cause an engine not to start. Technician B says incorrect valve timing may cause a power loss. Who is right?
    A. Technician A only
    B. Technician B only
    C. Both A and B
    D. Neither A nor B (A.12)

82. Technician A says timing specifications can be found on the underhood emission label. Technician B says the timing light can be connected to any plug wire to obtain the proper firing input for setting timing. Who is right?
    A. A only
    B. Technician B only
    C. Both A and B
    D. Neither A nor B (B.7)

83. Technician A says that nylon fuel line that is bent sharply causing a kink is allowable and will not affect fuel flow. Technician B says nylon fuel line cannot be repaired and the entire line must be replaced. Who is right?
    A. Technician A only
    B. Technician B only
    C. Both A and B
    D. Neither A nor B (C.3)

84. All of the following checks are correct for testing the coolant sensor and/or circuitry **EXCEPT:**
    A. resistance and voltage checks.
    B. DTCs and scan data.
    C. thermometer, heated water, and resistance check.
    D. diode check. (E.2 and E.4)

85. Technician A says carbon canister filters can be replaced. Technician B says filler caps with pressure and vacuum valves must be checked for contamination and damage. Who is right?
    A. Technician A only
    B. Technician B only
    C. Both A and B
    D. Neither A nor B (C.3 and D.4.3)

86. A mechanical fuel pump is being tested using a standard vacuum/pressure gauge at the fuel pump inlet. Technician A says this checks the condition of the diaphragm. Technician B says this checks the valve in the pump. Who is right?
    A. Technician A only
    B. Technician B only
    C. Both A and B
    D. Neither A nor B (C.4)

87. A faulty fuel pump is suspected. Which of these steps should the technician take first?
    A. Perform fuel pump pressure and volume tests.
    B. Check for fuel pump diagnostic trouble codes (DTC).
    C. Check the fuel filter.
    D. Check the fuel lines. (C.4)

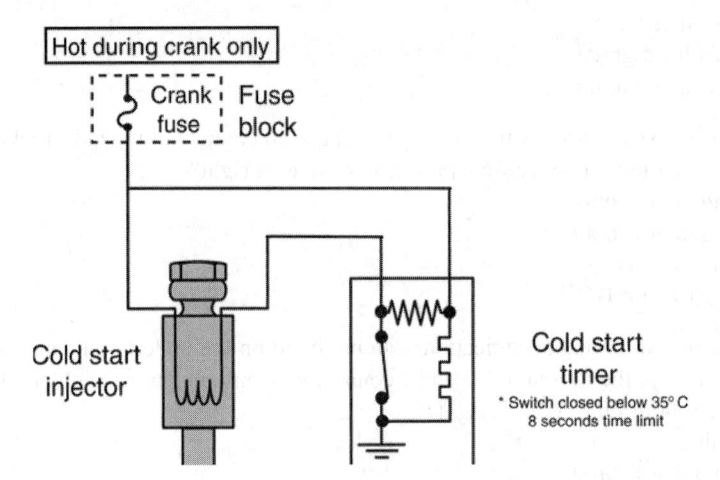

88. Referring to the figure, Technician A says the cold-start injector operates during engine cranking only. Technician B says coolant temperature determines how long the cold-start injector operates during cranking. Who is right?
    A. Technician A only
    B. Technician B only
    C. Both A and B
    D. Neither A nor B (C.8)

89. The first symptom of a restricted air filter is:
    A. loss of power.
    B. a no-start condition.
    C. excessive engine wear.
    D. excessive oil consumption. (C.9)

90. Technician A says when testing the starter control circuit, the voltage drop during cranking should not exceed 3.5 volts. Technician B says individual voltage drop tests on control circuit components may be necessary. Who is right?
    A. Technician A only
    B. B only
    C. Both A and B
    D. Neither A nor B (F.2.2)

91. All of the following could reduce turbocharger boost pressure **EXCEPT:**
    A. turbocharger bearing damage.
    B. wastegate stuck open.
    C. a leak between the turbo and throttle body.
    D. a stuck open engine thermostat. (C.15)

92. Replacement of an alternator is being discussed. Technician A says the alternator connectors should always be inspected for corrosion and/or distortion from overheating. Technician B says some replacement alternators come with a new connector. Who is right?
    A. Technician A only
    B. B only
    C. Both A and B
    D. Neither A nor B (F.3.3)

93. Technician A says if the positive crankcase ventilation (PCV) valve is stuck open, excessive airflow through the valve causes a rich air/fuel ratio. Technician B says if the PCV valve is restricted, excessive crankcase pressure forces blowby gases through the clean air hose into the air filter. Who is right?
    A. Technician A only
    B. B only
    C. Both A and B
    D. Neither A nor B (D.1.1)

94. A battery is being tested with a carbon pile. Technician A says to maintain the load for 30 seconds. Technician B says to apply a load equal to the cold crank rating of the battery. Who is right?
    A. Technician A only
    B. Technician B only
    C. Both A and B
    D. Neither A nor B (F.1.1)

95. Technician A says if the exhaust gas recirculation (EGR) valve remains open at idle and low speed, the idle will be rough. Technician B says if the EGR valve does not open, detonation can occur. Who is right?
    A. Technician A only
    B. Technician B only
    C. Both A and B
    D. Neither A nor B (D.2.1)

96. A battery is rated by all of the following means **EXCEPT:**
    A. cold-cranking amps.
    B. amp hour rating.
    C. reserve amps.
    D. reserve minutes. (F.1.1)

97. The LEAST-Likely cause of spark knock is:
    A. EGR valve stuck closed.
    B. fuel quality.
    C. carbon buildup on top of the pistons.
    D. EGR valve stuck open. (D.2.1)

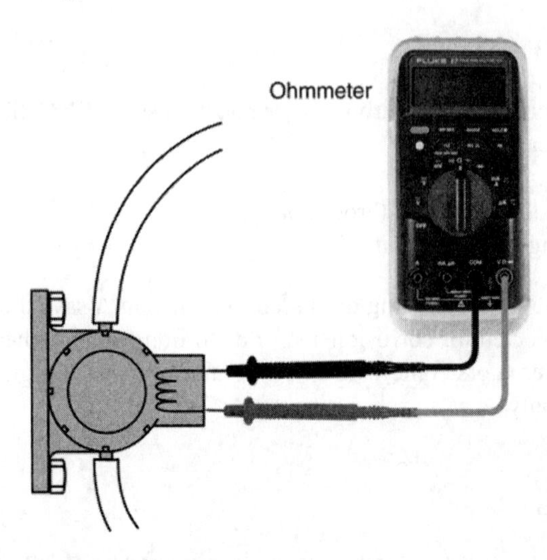

98. An EGR vacuum regulator solenoid (EGRV) is thought to be inoperative. Technician A says when an ohmmeter is connected, as shown, a lower than specified reading means the windings are open. Technician B says an infinite reading means the winding is shorted. Who is right?
    A. Technician A only
    B. Technician B only
    C. Both A and B
    D. Neither A nor B (D.2.3)

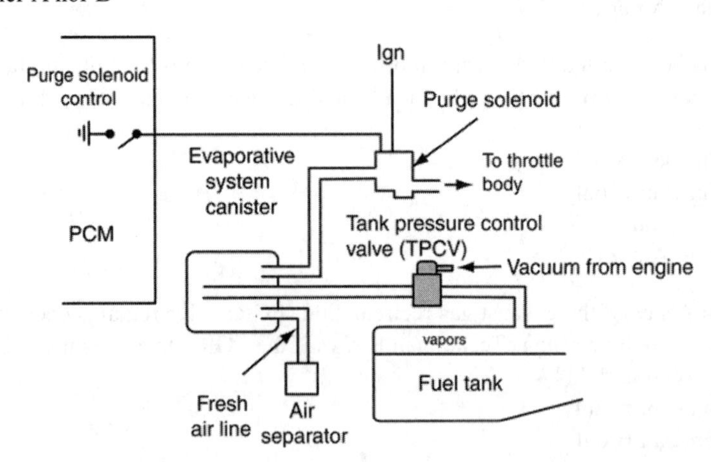

99. In the figure, Technician A says with a fuel tank pressure control valve off the vehicle, you could check the valve with a vacuum pump. Technician B says the valve must be installed on the vehicle, and a vacuum gauge must be used. Who is right?
    A. A only
    B. Technician B only
    C. Both A and B
    D. Neither A nor B (D.4.3)

100. Technician A says a diagnostic trouble code (DTC) tells you what component should be replaced. Technician B says if a fault exists that affects emissions, the emission reminder lamp will be illuminated. Who is right?
    A. Technician A only
    B. Technician B only
    C. Both A and B
    D. Neither A nor B (E.1 and E.2)

101. A mass airflow-controlled, port fuel-injected engine runs fine at idle but hesitates under acceleration with no DTCs stored. Technician A says to check for a restricted mass airflow sensor inlet screen. Technician B says dropout in the throttle position sensor signal could be the cause. Who is right?
    A. Technician A only
    B. Technician B only
    C. Both A and B
    D. Neither A nor B (E.3)

102. Technician A says the PCM will increase the fuel injector pulse width if there is excess oxygen in the exhaust. Technician B says if there is a lean condition, the SFT will show a minus value on the scan tool. Who is right?
    A. A only
    B. Technician B only
    C. Both A and B
    D. Neither A nor B (E.3)

103. Technician A says that a DMM/DVOM can be used to check an oxygen sensor. Technician B says that to check an oxygen sensor you need a diode tester. Who is right?
    A. A only
    B. Technician B only
    C. Both A and B
    D. Neither A nor B (E.4)

104. Technician A says an analog voltmeter cannot be used to check an O_2 sensor. Technician B says a test light can be used to check an O_2 sensor. Who is right?
    A. A only
    B. Technician B only
    C. Both A and B
    D. Neither A nor B (E.4)

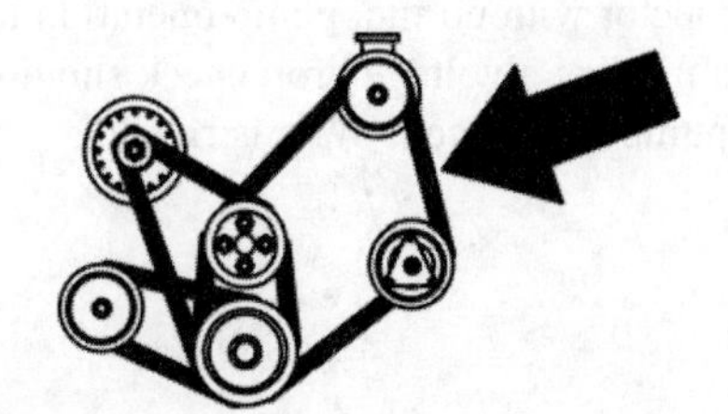

105. Referring to the figure, Technician A says a deteriorated belt could cause squealing and chirping. Technician B says belt deflection should be under 1/2 in (12.7 mm) per foot (30.5 cm) free span. Who is right?
    A. Technician A only
    B. Technician B only
    C. Both A and B
    D. Neither A nor B (F.3.2)

106. All of the following conditions could result in a voltage drop in a circuit **EXCEPT:**
   A. corrosion on connector terminals.
   B. cut strands in a multiple strand wire.
   C. using a heavier gauge wire.
   D. spread female terminals in a connector. (E.5)

107. Technician A says you will not harm the PCM with static electricity if the negative battery cable is disconnected. Technician B says you should never ground yourself to the vehicle while working on a PCM. Who is right?
   A. Technician A only
   B. Technician B only
   C. Both A and B
   D. Neither A nor B (E.7)

108. Technician A says some computer inputs are received from other computers. Technician B says one input might affect other computers. Who is right?
   A. Technician A only
   B. Technician B only
   C. Both A and B
   D. Neither A nor B (E.8)

109. The ignition module uses the digital signal received from the PCM for:
   A. rpm input.
   B. Hall effect timing.
   C. #1 cylinder signal.
   D. computed timing signal. (E.9)

110. A vehicle has a diagnostic trouble code (DTC) stored in memory. Technician A says after the vehicle is repaired, the DTCs should be cleared using a scan tool. Technician B says some vehicles require removing the PCM fuse to clear the DTC. Who is right?
   A. Technician A only
   B. Technician B only
   C. Both A and B
   D. Neither A nor B (E.10)

111. While testing a vehicle for no-start, Technician A says that since there is battery voltage at the electric fuel pump connector with no fuel pump operation it must be a bad fuel pump. Technician B disagrees stating that a voltage drop check should be made at the fuel pump connector before the fuel pump is replaced. Who is right?
   A. Technician A only
   B. B only
   C. Both A and B
   D. Neither A nor B (F.1.1)

112. A 12-volt battery has just failed a capacity test and is being charged at 40 amps. After three minutes of charge, with the charger still operating, a voltmeter is hooked up across the battery and reads 15.8 volts. What does this indicate?
   A. The battery should be slow charged and put back into service.
   B. The battery's electrolyte should be replaced.
   C. The battery is sulfated and should be replaced.
   D. This is normal; continue fast charge and return to service. (F.1.1)

113. A voltmeter is connected across a 12-volt battery. With the engine cranking, the voltmeter should not read less than:
    A. 12 volts.
    B. 10.5 volts.
    C. 9.6 volts.
    D. 7.5 volts. (F.1.1)

114. All of the following apply to OBD-II guidelines, **EXCEPT:**
    A. use of a standard list of diagnostic trouble codes.
    B. standard communication protocol.
    C. ability to record and store fault conditions when they occur.
    D. turn on the MIL if emission levels exceed four times the standards for that model year vehicle. (A.2, E.1 and E.2)

115. During cranking, Technician A says battery voltage should be no more than 5 volts. Technician B says cranking voltage doesn't matter, as long as the engine starts. Who is right?
    A. Technician A only
    B. Technician B only
    C. Both A and B
    D. Neither A nor B (F.1.1 and F.2.1)

116. When the belt and belt tension are satisfactory and the alternator output is low, Technician A says the alternator may be defective. Technician B says the problem could be high resistance in the alternator field circuit. Who is right?
    A. Technician A only
    B. Technician B only
    C. Both A and B
    D. Neither A nor B (F.3.1)

6 Additional Test Questions for Practice

Additional Test Questions

Please note the letter and number in parentheses following each question. They match the task in Section 4 that discusses the relevant subject matter. You may want to refer to the overview using the cross-referencing key to help with questions posing problems for you.

1. Technician A says that a DMM/DVOM can be used to check an oxygen sensor. Technician B says that to check an oxygen sensor you need a diode tester. Who is right?
   A. A only
   B. Technician B only
   C. Both A and B
   D. Neither A nor B (E.4)

2. All of the following symptoms are true of low fuel pressure **EXCEPT:**
   A. excessive sulfur smell.
   B. lack of power.
   C. engine surging.
   D. limited top speed. (C.1)

3. All of the following are true of installing a distributor **EXCEPT:**
   A. the crankshaft timing pointer must be aligned with the timing indicator.
   B. an indicator mark placed before removal aids in reassembly.
   C. on a four-cylinder engine, the distributor can be timed just after TDC on the #3 cylinder.
   D. the ignition rotor must point at the distributor cap terminal for the specified cylinder. (B.4 and B.7)

4. On an idling vehicle equipped with an AIR system, Technician A says the "MIL" will turn on after two minutes. Technician B says that the air helps cool the O_2 sensor. Who is right?
   A. Technician A only
   B. Technician B only
   C. Both A and B
   D. Neither A nor B (A.2 and D.3.1)

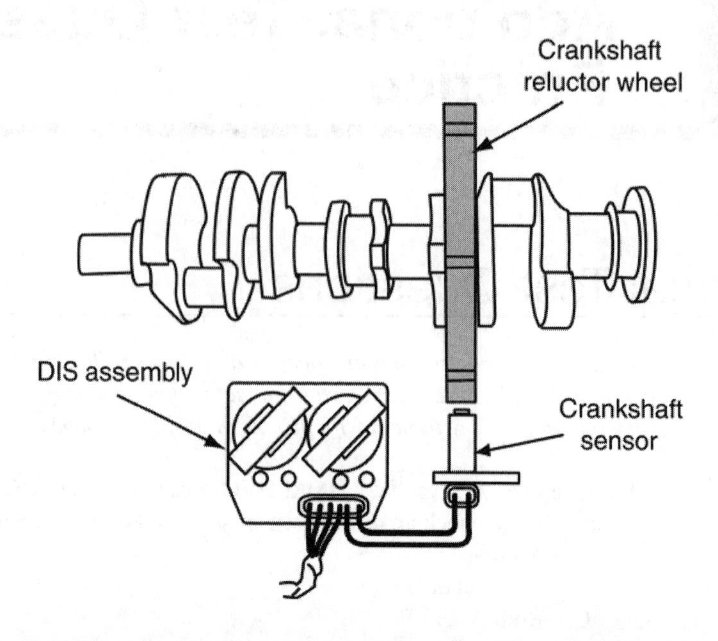

5. An engine equipped with a distributorless ignition system (DIS) as shown won't start. Technician A says a defective crankshaft position sensor could cause this. Technician B says an open ground wire to the DIS assembly could be the cause. Who is right?
   A. Technician A only
   B. Technician B only
   C. Both A and B
   D. Neither A nor B (B.1)

6. The ignition module uses a digital signal from the PCM for:
   A. timing advance control.
   B. #1 TDC signal.
   C. rpm input.
   D. fuel economy information. (A.2 and B.7)

7. All statements are true about adjusting valves **EXCEPT:**
   A. mechanical lifters are adjusted by rotating the nut on the rocker arm until the specified clearance is achieved.
   B. after hydraulic lifters are pumped up, adjust by rotating the nut until the specified clearance is achieved.
   C. use a feeler gauge for mechanical lifters.
   D. the piston should be at top dead center. (A.11)

8. Which of these would cause a double knocking noise with the engine at an idle?
   A. Worn piston wrist pins.
   B. Excessive timing chain deflection.
   C. Worn main bearing.
   D. Excessive main bearing thrust clearance. (A.3)

9. Technician A says secondary air-injection systems must be monitored for proper operation on a vehicle certified as OBD-II compliant. Technician B says secondary air-injection systems are not used on OBD-II compliant vehicles. Who is right?
   A. Technician A only
   B. Technician B only
   C. Both A and B
   D. Neither A nor B (A.2)

10. A technician is performing a compression test. Which statement below is LEAST Likely true?
    A. All cylinders with higher than normal readings could be caused by carbon buildup.
    B. All cylinders reading even, but lower than normal, may be caused by a slipped timing chain.
    C. Low readings on two adjacent cylinders may be caused by a blown head gasket.
    D. A low reading on one cylinder may be caused by a vacuum leak at that cylinder. (A.7)

11. Technician A says an analog voltmeter cannot be used to check an O_2 sensor. Technician B says a test light can be used to check an O_2 sensor. Who is right?
    A. A only
    B. Technician B only
    C. Both A and B
    D. Neither A nor B (E.4)

12. During a cylinder leakdown test on a four-cylinder engine, air is heard coming from the #3 spark plug hole as cylinder #4 is being checked. Technician A says that this could be caused by a blown head gasket. Technician B says this could be caused by a cracked engine block. Who is right?
    A. Technician A only
    B. Technician B only
    C. Both A and B
    D. Neither A nor B (A.8)

13. A four-cylinder engine is making a loud metallic knocking that gets louder as the engine warms up or if the throttle is quickly snapped open. The noise almost disappears when the spark for cylinder # 3 is shorted to ground. Technician A says the problem is most likely a cracked flywheel. Technician B says the problem is most likely a loose connecting rod bearing. Who is right?
    A. Technician A only
    B. Technician B only
    C. Both A and B
    D. Neither A nor B (A.3)

14. A multi-trace oscilloscope can test all of the following **EXCEPT:**
    A. air-fuel ratio.
    B. manifold absolute pressure sensor.
    C. throttle position sensor.
    D. crank position sensor. (A.9)

15. Technician A says that overfilling the crankcase with oil can cause hydraulic lifter noise due to oil aeration. Technician B says excessive oil pressure from using motor oil with a viscosity rating that is too high can cause hydraulic lifter noise. Who is right?
    A. Technician A only
    B. Technician B only
    C. Both A and B
    D. Neither A nor B (A.3)

16. Which of the following would MOST Likely cause weak spark at the spark plug wires?
    A. Timing advance out of spec
    B. Leaking secondary insulation
    C. Low resistance in the secondary circuit
    D. High resistance in the primary circuit (B.1 and B.3)

17. A restricted exhaust will cause vacuum readings to:
    A. drop off about 3 inches at an idle.
    B. drop off about 8 inches at an idle.
    C. fluctuate between 16 and 21 inches at an idle.
    D. show a continuous gradual drop as engine speed is increased. (A.5 and C.14)

18. The LEAST-Likely cause of blue exhaust smoke is:
    A. worn cylinder walls.
    B. worn valve seals.
    C. stuck piston rings.
    D. worn valve seats. (A.4)

19. All of the following are true of the cylinder leakage test **EXCEPT:**
    A. air loss and bubbles in the radiator indicate a bad head gasket or engine casting crack.
    B. air loss from the oil filler cap indicates worn piston rings.
    C. a gauge reading of 100% indicates no cylinder leakage.
    D. air loss from the exhaust indicates a valve problem. (A.8)

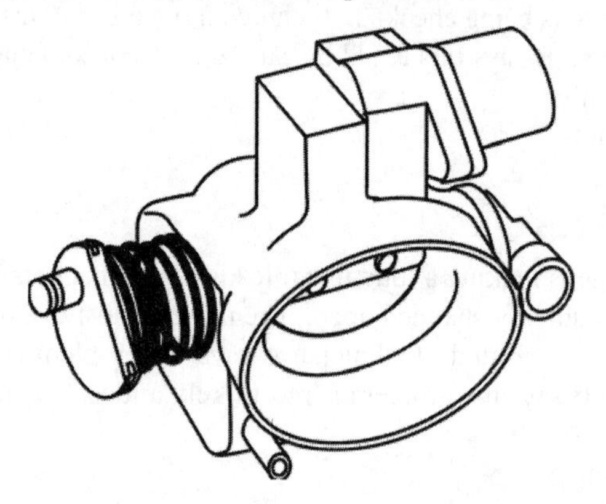

20. Technician A says the throttle body shown in the figure must be removed to be cleaned. Technician B says the minimum airflow rate and throttle plate angle needs to be checked and adjusted if necessary after cleaning. Who is right?
    A. Technician A only
    B. B only
    C. Both A and B
    D. Neither A nor B (C.7 and C.11)

21. While performing a scan test on an OBD-II certified vehicle, a diagnostic trouble code P1336 is retrieved. Technician A says that a first digit of P means the code is a generic trouble code. Technician B says that a second digit of 1 means the code is a manufacturer specific code. Who is right?
    A. Technician A only
    B. B only
    C. Both A and B
    D. Neither A nor B (E.1)

22. During a cylinder power balance test, there is no rpm drop on cylinder #3. Technician A says that the cylinder is not contributing. Technician B says that the cylinder may have an inoperative spark plug. Who is right?
    A. Technician A only
    B. Technician B only
    C. Both A and B
    D. Neither A nor B (A.6 and B.1)

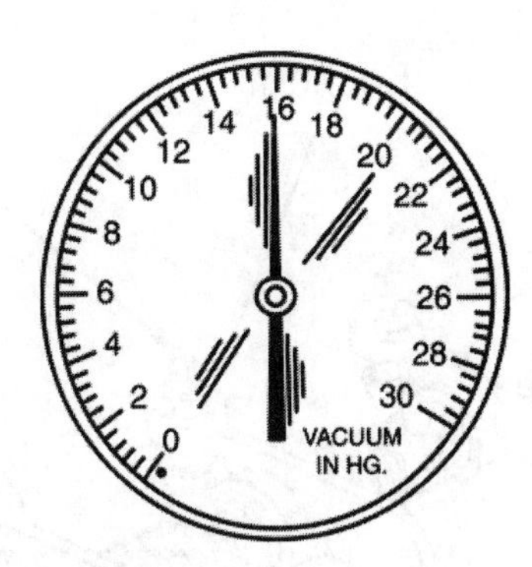

23. The vacuum gauge shows a fluctuating motion from 15 to 21 in. Hg at idle. Technician A says this could be caused by a loose exhaust manifold. Technician B says this could be caused by a burned valve. Who is right?
    A. Technician A only
    B. Technician B only
    C. Both A and B
    D. Neither A nor B (A.5)

24. A vacuum-operated fuel pressure regulator is used on port-style fuel injection for which of the following reasons?
    A. To increase fuel delivery under high load conditions
    B. To prevent fuel-pressure leak-down when the engine is turned off
    C. To provide a constant pressure drop across the injector due to changing manifold pressure
    D. To improve injector spray patterns (A.2 and C.6)

25. Two technicians are discussing electrical symbols Technician A says that all ground symbols are the same on any electrical schematic. Technician B says that there is more than one symbol for a ground. Who is right?
    A. Technician A only
    B. Technician B only
    C. Both A and B
    D. Neither A nor B (A.9 and C.8)

26. A turbocharger requires frequent replacement due to bearing failure. Technician A says oil contamination from sludge buildup in the engine could be the cause. Technician B says restricted oil passages to the turbocharger could be the problem. Who is right?
    A. Technician A only
    B. Technician B only
    C. Both A and B
    D. Neither A nor B (C.15)

27. A vehicle with an electronic ignition fails to start. Technician A says this could be caused by a defective crankshaft sensor connection. Technician B says this could be caused by a defective ignition module. Who is right?
    A. Technician A only
    B. Technician B only
    C. Both A and B
    D. Neither A nor B (B.1 and B.8)

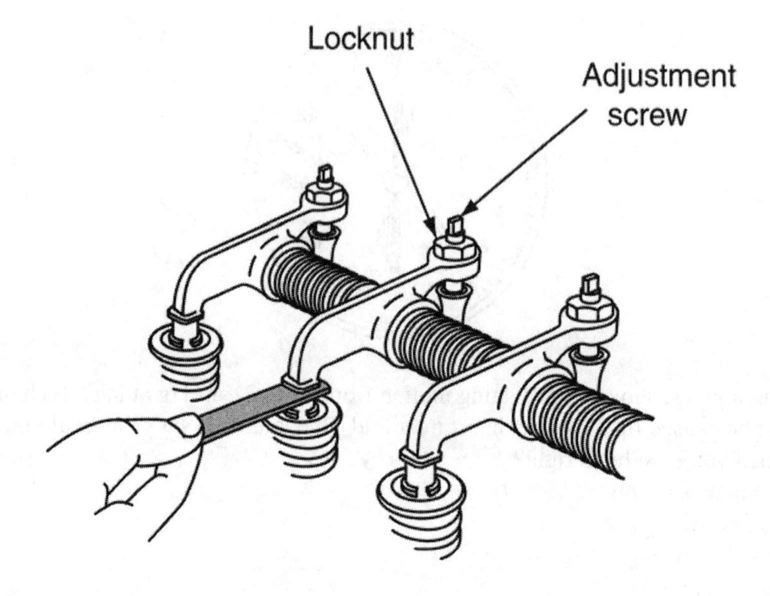

28. If the measurement in the figure is set too wide, Technician A says it will retard valve timing. Technician B says it will reduce valve overlap. Who is right?
    A. Technician A only
    B. Technician B only
    C. Both A and B
    D. Neither A nor B (A.11)

29. The customer complains of sluggish performance and poor fuel economy. The vehicle has a DTC for O_2 voltage low/lean exhaust and has high CO exhaust emissions. The O_2 sensor is tested separately and is functioning properly. Technician A says the fuel injectors may be leaking. Technician B says secondary air may be diverted upstream after closed-loop status. Who is right?
    A. Technician A only
    B. Technician B only
    C. Both A and B
    D. Neither A nor B (A.10, C.1, D.3.1, D.3.2, E.1, E.2 and E.3)

30. A four-gas exhaust emissions analyzer may be used to help diagnose all of the following problems **EXCEPT:**
    A. slow or lazy oxygen sensor.
    B. engine cylinder misfire.
    C. defective catalytic converter.
    D. plugged or restricted fuel injector. (A.10)

31. A scan tool is being used in addition to a thermometer to check thermostat operation. Technician A says using both tools will aid in making an accurate diagnosis of thermostat operation. Technician B says thermostat operation can also be checked visually by running the engine until it is hot. Who is right?
    A. A only
    B. Technician B only
    C. Both A and B
    D. Neither A nor B (A.13)

32. While performing a valve adjustment, Technician A says the crankshaft must be placed in the proper position so that the valve lifter is riding on the camshaft lobe. Technician B says that adjusting valves with too little clearance may cause rough running and burnt valves. Who is right?
    A. Technician A only
    B. Technician B only
    C. Both A and B
    D. Neither A nor B (A.11)

33. Many cooling systems use a thermal control fan drive coupling (fan clutch). Technician A says the thermostatic coil controls the opening and closing of the orifice inside the coupling. Technician B says when the thermostatic coil is cold, the orifice is open. Who is right?
    A. A only
    B. Technician B only
    C. Both A and B
    D. Neither A nor B (A.14)

34. During engine testing a technician observes tailpipe CO readings that are higher than normal at idle. Which of these is the MOST-Likely cause?
    A. Constant high-oxygen-sensor voltage signal
    B. An EGR valve not seating
    C. A bad ignition module
    D. An EVAP purge solenoid stuck open (A.10, D.2.1 and D.4.1)

35. A slipped timing belt can cause all **EXCEPT:**
    A. poor fuel mileage.
    B. no-start.
    C. high manifold vacuum.
    D. low power. (A.12)

36. High hydrocarbon (HC) emissions may be caused by all the following **EXCEPT:**
    A. a cylinder misfire.
    B. an excessive lean condition.
    C. a faulty fuel-pressure regulator.
    D. power relay. (A.10)

37. When a vacuum leak is suspected for a rough idle complaint, the best check would be to use:
    A. carburetor cleaner.
    B. propane.
    C. water.
    D. a stethoscope. (C.10)

38. With a thermometer taped to the upper radiator hose as shown above, and the vehicle fully warmed up, the temperature indication should be:
    A. ambient temperature.
    B. within a few degrees of the engine thermostat temperature rating.
    C. one-half of the engine thermostat temperature rating.
    D. less than the temperature measured at the lower radiator hose. (A.13)

39. All of the following conditions can cause high current readings during a starter current draw test **EXCEPT:**
    A. shorted armature windings.
    B. over-advanced ignition timing.
    C. corroded battery cables.
    D. insufficient crankshaft bearing clearances. (F.2.1)

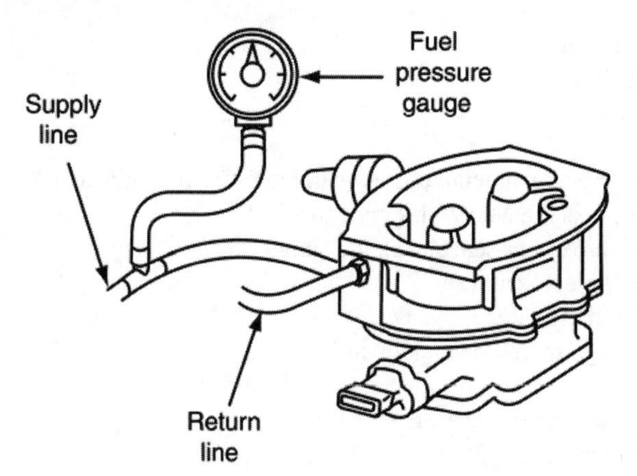

40. In the figure, Technician A says the throttle body must be completely disassembled before soaking it in cleaning solution. Technician B says the test being performed is for fuel pressure. Who is right?
    A. Technician A only
    B. Technician B only
    C. Both A and B
    D. Neither A nor B (C.6 and C.7)

41. While applying dielectric silicone to an ignition module mounting surface, Technician A says dielectric silicone is used to dissipate heat. Technician B says without dielectric silicone, the ignition module may experience overheat conditions. Who is right?
    A. Technician A only
    B. Technician B only
    C. Both A and B
    D. Neither A nor B (B.9)

42. Technician A says that some resistance in the power distribution circuits may harm the ignition switch or starter due to low voltage. Technician B says the power to the ignition switch is never fused. Who is right?
    A. Technician A only
    B. Technician B only
    C. Both A and B
    D. Neither A nor B (B.3 and E.6)

43. When discussing pickup coil testing, all of the following are true **EXCEPT:**
    A. each pickup coil should be within the manufacturer's specified resistance value.
    B. a resistance reading below manufacturer's specification indicates an open pickup coil.
    C. a resistance reading above manufacturer's specification indicates an open pickup coil.
    D. an erratic reading while wiggling the pickup coil wires indicates that the pickup coil is intermittent. (B.8)

44. Technician A says the ignition module may control ignition coil dwell time on some ignition systems. Technician B says the ignition module may control ignition coil current flow on some ignition systems. Who is right?
    A. Technician A only
    B. Technician B only
    C. Both A and B
    D. Neither A nor B (B.9)

45. The LEAST-Likely cause of ignition coil failure is:
    A. prolonged open circuit in the secondary.
    B. overheating the coil.
    C. coil case cracking.
    D. open ignition module primary circuit. (B.1 and B.6)

46. Technician A says oil accumulation in the air filter housing can be an indication of a plugged PCV system. Technician B says that oil accumulation in the air filter housing can be an indication of excessive engine blow-by from a worn out engine. Who is right?
    A. Technician A only
    B. Technician B only
    C. Both A and B
    D. Neither A nor B (D.1.1)

47. Technician A says that the ignition module controls timing during starting on most ignition systems. Technician B says that on some late model systems, the powertrain control module (PCM) has full control of timing at all times. Who is right?
    A. Technician A only
    B. Technician B only
    C. Both A and B
    D. Neither A nor B (B.9)

48. Technician A says the camshaft sensor may be rotated to adjust the base ignition timing. Technician B says on some systems, the crankshaft sensor may be moved to obtain proper clearance between the sensor and interrupter blades. Who is right?
    A. Technician A only
    B. B only
    C. Both A and B
    D. Neither A nor B (B.8)

49. An MFI vehicle stalls intermittently at idle and has low long-term fuel trim correction values stored when checked with a scan tool. All of the following conditions could cause this **EXCEPT:**
    A. leaking fuel injectors.
    B. unmetered air leaking into the engine.
    C. a leaking fuel-pressure regulator diaphragm.
    D. a fuel-pressure regulator sticking closed. (C.1, C.6 and C.10)

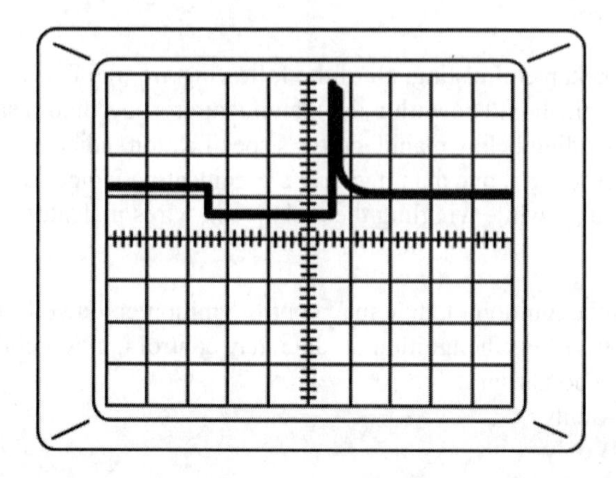

50. Referring to the sample pattern, a vehicle with port fuel injection is running roughly. A labscope shows each injector waveform to be identical, except for one that has a considerably shorter voltage spike than the others. Which of the following is the MOST-Likely cause?
    A. Bad PCM
    B. Open connection at the injector
    C. Shorted injector winding
    D. Low charging system voltage (A.9 and C.1)

51. Technician A says a greenish white corrosion on terminals results in high resistance. Technician B says loose retaining nuts cause high resistance in connector ring terminals. Who is right?
    A. Technician A only
    B. Technician B only
    C. Both A and B
    D. Neither A nor B (E.5)

52. Technician A says ignition timing is usually set with the engine running and the computer power feed disconnected. Technician B says ignition timing is usually set with the engine running at 2,500 rpm. Who is right?
    A. Technician A only
    B. Technician B only
    C. Both A and B
    D. Neither A nor B (B.7)

53. Technician A says if you are testing an ignition related no-start problem, you should always check for available spark at an ignition wire first. Technician B says if you have a test light connected to the negative side of the coil while cranking and the test light flickers, you need to test the secondary ignition system. Who is right?
    A. Technician A only
    B. Technician B only
    C. Both A and B
    D. Neither A nor B (B.1)

54. A faulty fuel pump is suspected. Which of these steps should the technician take first?
    A. Perform fuel pump pressure and volume tests.
    B. Check for fuel pump diagnostic trouble codes (DTC).
    C. Check the fuel filter.
    D. Check the fuel lines. (C.4)

55. When voltage drop testing the power and ground distribution circuits, Technician A says all connections must be clean and free of corrosion. Technician B says corrosion adds unwanted resistance. Who is right?
    A. Technician A only
    B. B only
    C. Both A and B
    D. Neither A nor B (E.5)

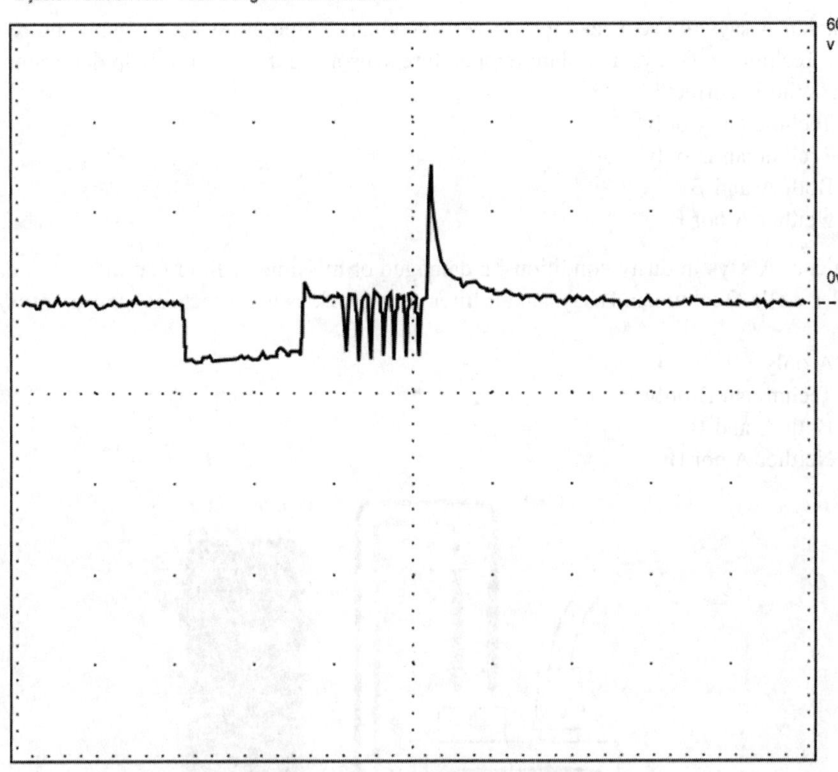

56. Two technicians are discussing the injector waveform in the figure. Technician A says the waveform indicates the injector is power switched. Technician B says the waveform indicates the injector is ground switched. Who is right?
    A. Technician A only
    B. Technician B only
    C. Both A and B
    D. Neither A nor B (D.2.1 and D.2.2)

57. Technician A says that an oscilloscope can be used to watch the O2 sensor signal switch from rich to lean status. Technician B says fuel injectors can only be tested using the fuel-injector balance tester. Who is right?
    A. A only
    B. Technician B only
    C. Both A and B
    D. Neither A nor B (C.8 and E.4)

58. Technician A says a defective MAP sensor may cause a rich or lean air/fuel ratio. Technician B says by measuring the MAP signal under different engine operating conditions, the MAP sensor can be accurately diagnosed. Who is right?
    A. A only
    B. Technician B only
    C. Both A and B
    D. Neither A nor B (C.1 and E.4)

59. Technician A says to use a scan tool to verify coolant temperature sensor input and related DTCs. Technician B says a coolant temperature sensor input is used to help determine loop status. Who is correct?
    A. Technician A only
    B. Technician B only
    C. Both A and B
    D. Neither A nor B (E.2 and E.4)

60. Technician A says in dusty conditions, a damaged or missing air filter can increase wear on cylinder walls. Technician B says an air filter problem does not affect fuel consumption. Who is right?
    A. A only
    B. Technician B only
    C. Both A and B
    D. Neither A nor B (C.9)

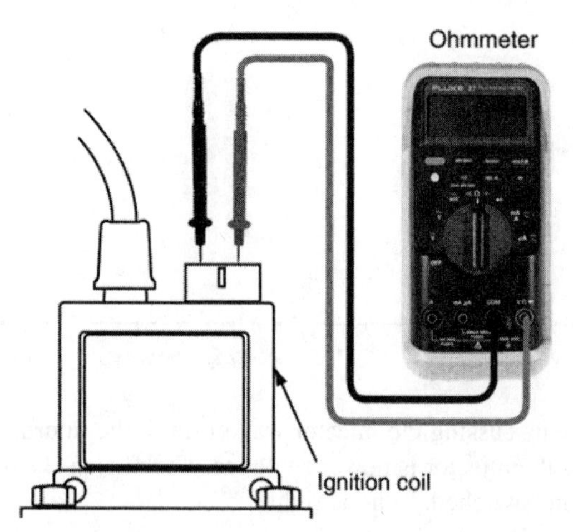

61. Technician A says the ohmmeter in the figure reads infinity and means the circuit has little or no resistance. Technician B says this means a circuit is open. Who is right?
    A. Technician A only
    B. B only
    C. Both A and B
    D. Neither A nor B (B.6)

62. Technician A says installing a rebuilt alternator in a vehicle with a weak battery without recharging the battery may cause premature alternator failure. Technician B says a battery open circuit voltage reading of 12.6 volts indicates the battery is good and fit for service. Who is right?
    A. A only
    B. Technician B only
    C. Both A and B
    D. Neither A nor B (F.1.1 and F.3.1)

63. While discussing the fuel pump electrical control circuits, Technician A says some vehicles use a fuel pump relay to energize the fuel pump. Technician B says some vehicles will not energize the fuel pump without a reference pulse from the crank sensor. Who is right?
    A. Technician A only
    B. Technician B only
    C. Both A and B
    D. Neither A nor B (D.3.1 and D.3.3)

64. A turbocharged engine is experiencing excessive oil consumption and blue smoke from the tailpipe. This could be caused by:
    A. a dirty oil filter.
    B. worn compressor shaft bearings.
    C. a dirty air filter.
    D. a broken exhaust pipe. (C.15)

65. A vehicle is equipped with a vented gas cap. Technician A says if a nonvented cap is installed in this vehicle, the gas tank could collapse. Technician B says that if a nonvented cap is installed, the vehicle could be starved for fuel at high speeds. Who is right?
    A. Technician A only
    B. Technician B only
    C. Both A and B
    D. Neither A nor B (C.3)

66. A carbureted vehicle has a tip-in hesitation when warm. Which of these should be checked first?
    A. Stuck open choke
    B. Accelerator pump
    C. Fuel filter
    D. Fuel return line (C.1)

67. Technician A says that prior to adjusting the throttle plates on a multiport injected engine throttle body, the throttle plates and bore should be cleaned if dirty or varnished. Technician B says the throttle position sensor needs to be readjusted on some vehicles after the throttle plate angle adjustment is complete. Who is right?
    A. Technician A only
    B. Technician B only
    C. Both A and B
    D. Neither A nor B (C.7 and C.11)

68. A multiport fuel-injected vehicle has poor fuel economy, yet starts and runs fine. Technician A says the fuel return line may be restricted. Technician B says the fuel-pressure regulator may be stuck. Who is right?
    A. Technician A only
    B. Technician B only
    C. Both A and B
    D. Neither A nor B

(C.6)

69. Technician A says oil found inside the air filter housing may be caused by a plugged or restricted PCV valve or hose. Technician B says a plugged PCV valve will cause excessive crankcase blow-by. Who is right?
    A. A only
    B. Technician B only
    C. Both A and B
    D. Neither A nor B

(D.1.1 and D.1.2)

70. Low battery or system voltage can cause all of the following, **EXCEPT** increased:
    A. ignition dwell.
    B. injector on-time.
    C. idle speed.
    D. steering effort.

(E.3 and F.1.1)

71. Technician A says that for certain common DTCs, you can replace the component without using the flow chart. Technician B says you should always use the flow chart, but certain steps in the flow chart can be skipped. Who is right?
    A. Technician A only
    B. Technician B only
    C. Both A and B
    D. Neither A nor B

(E.2 and E.10)

72. Technician A says because of extreme temperatures in the exhaust stream, restricted EGR passages are not a problem. Technician B says that before any EGR parts are replaced, all passages should be clean. Who is right?
    A. Technician A only
    B. B only
    C. Both A and B
    D. Neither A nor B

(D.2.3)

73. All of the following statements about PCM inputs are true **EXCEPT:**
    A. digital voltmeters/ohmmeters may be used for diagnosis.
    B. some inputs, but not all, are very low voltage.
    C. with practice, an experienced technician can use an analog meter.
    D. the O_2 sensor produces very low voltage.

(E.5)

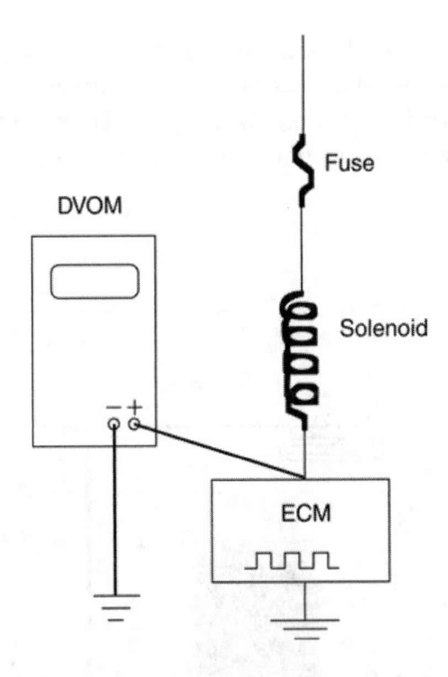

74. In the diagram above, Technician A says the meter will read 12 volts when the solenoid is commanded off by the ECM. Technician B says when the solenoid turns on, the voltmeter will read 12 volts. Who is right?
   A. A only
   B. Technician B only
   C. Both A and B
   D. Neither A nor B (E.5)

75. The LEAST-Likely symptom resulting from an evaporative-emission system failure is:
   A. increased tailpipe emissions.
   B. low vehicle emissions.
   C. malfunction indicator lamp (MIL) illumination.
   D. fuel odor. (D.4.1 and D.4.2)

76. Technician A says that diagnostic trouble codes (DTCs) can be retrieved by using a scan tool. Technician B says in some cases DTCs can be retrieved using an analog voltmeter. Who is right?
   A. Technician A only
   B. Technician B only
   C. Both A and B
   D. Neither A nor B (E.1)

77. Technician A says when high resistance is found in a circuit, check for burned wires, connector ring terminals, loose retaining nuts, or other wire and connector concerns. Technician B says that greenish-white corrosion can happen at any point where the wire insulation has been pierced or opened in any way. Who is right?
   A. Technician A only
   B. Technician B only
   C. Both A and B
   D. Neither A nor B (E.5)

78. Technician A says that when checking a pulsed secondary air-injection system, air pressure pulses felt at the fresh air intake hose indicate a bad aspirator or reed valve. Technician B says that when testing secondary air-injection systems, the carbon monoxide (CO) reading on a gas analyzer can confirm normal air-injection operation. Who is right?
    A. A only
    B. Technician B only
    C. Both A and B
    D. Neither A nor B (D.3.1 and D.3.3)

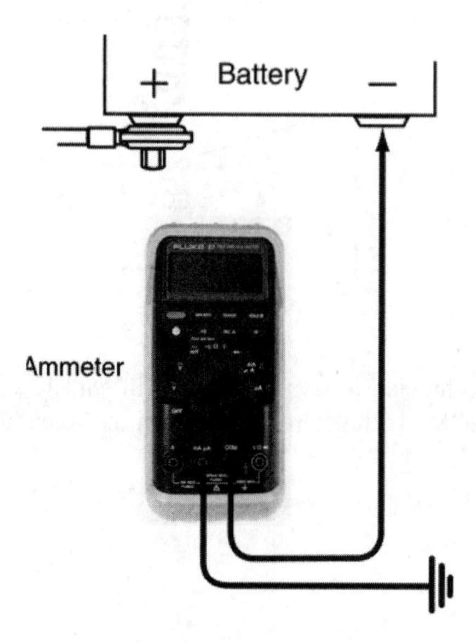

79. An ammeter, set in the milliamp position, is connected in series between the negative battery cable and ground, as shown. What is being measured?
    A. Starter draw
    B. Battery drain
    C. Regulated voltage
    D. Voltage drop (F.1.1)

80. Which of these is a function of an AIR system diverter valve?
    A. Prevent backfire on deceleration
    B. Enrich fuel mixture on deceleration
    C. Divert cold air into the passenger compartment
    D. Turn the A/C compressor off (D.3.3)

81. An EFI engine has poor acceleration when the vehicle is suddenly accelerated to WOT. Idle and cruise performance are fine. Technician A says a faulty mass airflow sensor could cause this. Technician B says faulty WOT position switch contacts could cause this. Who is right?
    A. Technician A only
    B. B only
    C. Both A and B
    D. Neither A nor B (C.1 and E.4)

82. Technician A says a scan tool is required to retrieve trouble codes. Technician B says OBD-II compliant vehicles use a standardized trouble code format. Who is right?
    A. Technician A only
    B. Technician B only
    C. Both A and B
    D. Neither A nor B (E.1)

83. Technician A says an analog voltmeter should be used for testing O2 sensor voltage output. Technician B says analog voltmeters are typically of low input impedance design and cannot be used because they draw too much current and will give inaccurate readings. Who is right?
    A. Technician A only
    B. B only
    C. Both A and B
    D. Neither A nor B (E.4)

84. An evaporative system diagnostic trouble code (DTC) may be set by all of the following **EXCEPT:**
    A. injector failure.
    B. powertrain control module (PCM).
    C. vacuum leak.
    D. solenoid failure. (D.4.1 and D.4.2)

85. A primary ignition circuit on a vehicle checks good, but there is no spark from the coil wire. This could be caused by a(n):
    A. defective coil.
    B. grounded rotor.
    C. overheated transistor.
    D. open diode. (B.6)

86. Technician A says a key-off current draw on a battery is called parasitic load and should not exceed .05A amps for most cars. Technician B says the alternator supplies the current for vehicle electrical loads once the engine is running under most operating conditions. Who is right?
    A. Technician A only
    B. Technician B only
    C. Both A and B
    D. Neither A nor B (F.1.1)

87. Technician A says today's computer systems have the ability to communicate between multiple computers. Technician B says one input might affect several computers. Who is right?
    A. Technician A only
    B. Technician B only
    C. Both A and B
    D. Neither A nor B (E.8)

88. A multiport fuel-injected vehicle has a rough extended idle. Technician A says this could be caused by a cracked or disconnected hose between the fuel tank and the EVAP canister. Technician B says a malfunctioning EVAP system can cause idle problems. Who is right?
    A. Technician A only
    B. B only
    C. Both A and B
    D. Neither A nor B (D.4.1 and D.4.3)

89. All of these will cause zero battery charging from an alternator **EXCEPT:**
    A. open field circuit.
    B. burned out or broken fusible link.
    C. broken belt.
    D. one open diode. (F.3.1)

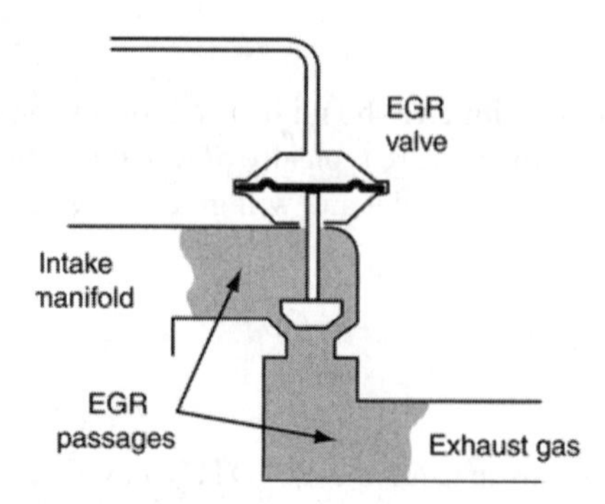

90. Technician A says to always check the exhaust passages when replacing an exhaust gas recirculation (EGR) valve, as shown in the figure. Technician B says an EGR valve that tests bad is inoperative; no other checks are needed. Who is right?
    A. A only
    B. Technician B only
    C. Both A and B
    D. Neither A nor B (D.2.3)

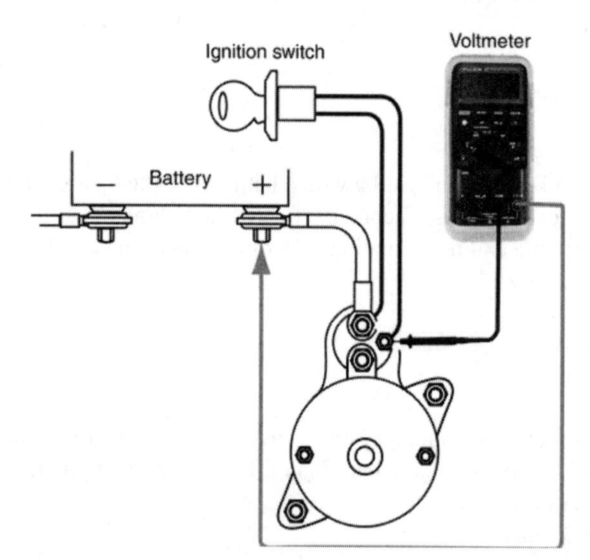

91. A voltage-drop test of the starter motor control circuit is being performed as shown in the figure. With the ignition disabled, the engine is cranked over with the voltmeter on the lowest scale. Technician A says that a reading below 2.5 volts is OK. Technician B says that if a reading above 1 volt is obtained, individual components in the circuit will need to be tested to find the problem. Who is right?
    A. Technician A only
    B. B only
    C. Both A and B
    D. Neither A nor B (E.5 and F.2.2)

92. Technician A says to check for battery drain, install a voltmeter from the positive battery terminal to the negative battery terminal. Technician B says to install an ohmmeter from the negative battery cable to a known good ground. Who is right?
    A. Technician A only
    B. Technician B only
    C. Both A nor B
    D. Neither A nor B (F.1.1)

93. Technician A says that if some EGR passages are plugged in the intake manifold, the engine may misfire when the EGR valve opens due to excessive EGR flow to the remaining cylinders. Technician B says that plugged EGR passages may cause a failed IM240 emissions test for high oxides of nitrogen. Who is right?
    A. Technician A only
    B. Technician B only
    C. Both A and B
    D. Neither A nor B (D.2.1)

94. A battery has been on a slow charger and the specific gravity is 1.15. Technician A says the battery is completely charged. Technician B says the electrolyte temperature should not exceed 125°F while recharging the battery. Who is right?
    A. Technician A only
    B. B only
    C. Both A and B
    D. Neither A nor B (F.1.1)

95. Technician A says you can test for a parasitic battery drain with a battery tester. Technician B says you must connect an ammeter in series to test for parasitic battery drain. Who is right?
    A. Technician A only
    B. B only
    C. Both A and B
    D. Neither A nor B (F.1.1)

96. High current draw and low cranking speed usually indicate:
    A. excessive resistance in the starter circuit.
    B. a defective battery.
    C. a defective starter.
    D. a defective ignition switch. (F.2.1)

97. A vehicle is hard to start due to low cranking speed. A starter draw test reveals excessive current draw from the starter. Technician A says the battery cables should be voltage-drop tested to determine if there is excessive resistance. Technician B says the starter motor armature may be dragging on the field coils. Who is right?
    A. Technician A only
    B. B only
    C. Both A and B
    D. Neither A nor B (F.2.1 and F.2.2)

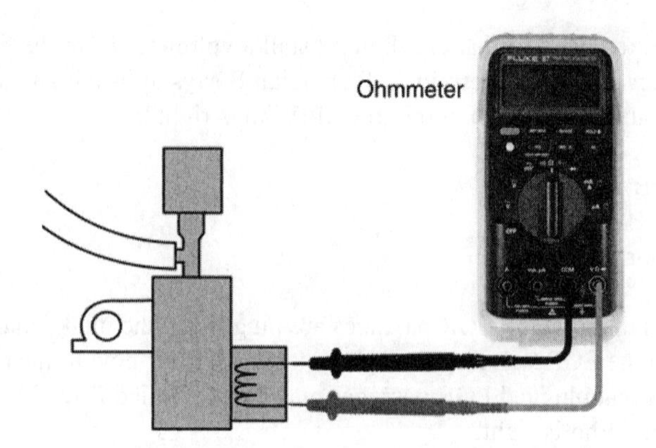

98. A diagnostic trouble code for an EGR problem is retrieved in the figure. A reading of infinite between terminals of the EGR vacuum regulator could mean:
    A. nothing.
    B. the regulator is defective.
    C. the EGR valve is defective.
    D. the MAP sensor is defective. (D.2.1 and D.2.2)

99. When testing voltage drop while cranking the engine with the ignition disabled, Technician A says the voltage drop on the positive battery cable should be no more than 5 volts. Technician B says the voltage drop on the positive battery cable should be no more than 1 volt. Who is right?
    A. Technician A only
    B. Technician B only
    C. Both A and B
    D. Neither A nor B (F.2.2)

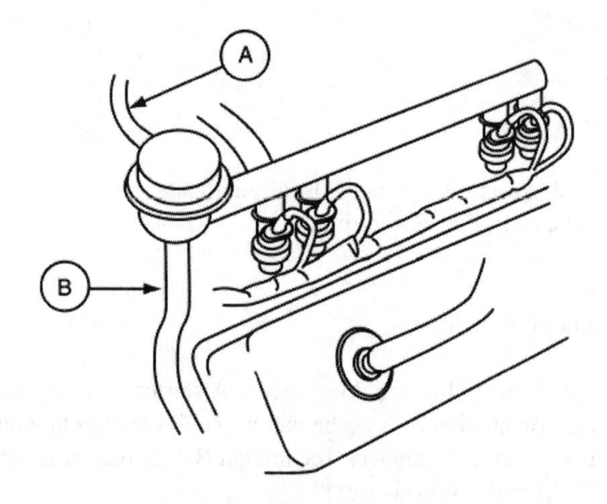

100. Technician A says by disconnecting the vacuum line to the pressure regulator during idle, as shown at A in the figure, a pressure increase should be noted on the fuel gauge. Technician B says if fuel pressure stops dropping after the return fuel line is restricted at B while doing a key-on/engine-off fuel-pressure test, the pressure regulator is stuck open. Who is right?
    A. Technician A only
    B. Technician B only
    C. Both A and B
    D. Neither A nor B (C.6)

101. A technician connects a jumper wire between the alternator B + and F terminals during a field circuit and alternator test. Technician A says if this corrects a low voltage reading, the wiring harness from the alternator to the regulator could be faulty. Technician B says this bypasses the voltage regulator. Who is right?
    A. Technician A only
    B. Technician B only
    C. Both A and B
    D. Neither A nor B (F.3.1 and F.3.2)

102. Technician A says an air-injection system directs air pump output to the catalytic converter during engine warm-up and switches air to the exhaust manifold during closed loop operation. Technician B says air injection will have no effect on tailpipe oxygen readings. Who is right?
    A. Technician A only
    B. Technician B only
    C. Both A and B
    D. Neither A nor B (D.3.3)

103. Technician A says to prevent backfiring during deceleration on pump-driven air injection systems, a diverter valve is used. Technician B says to prevent exhaust gases from back flowing into the air-injection control valves or pump, check valves are used in the exhaust manifold and converter feed pipes. Who is right?
    A. Technician A only
    B. Technician B only
    C. Both A and B
    D. Neither A nor B (D.3.1)

104. A vehicle emits a belt squeal when starting and on acceleration. Technician A says the alternator bearings may be defective. Technician B says that the alternator belt may be loose. Who is right?
    A. Technician A only
    B. B only
    C. Both A and B
    D. Neither A nor B (F.3.2)

105. All of the following are true about the complaint verification process **EXCEPT:**
    A. isolate the system causing the complaint.
    B. identify conditions under which the concern occurs.
    C. road test.
    D. review the repair order. (A.1)

7 Appendices

Answers to the Test Questions for the Sample Test Section 5

1. C
2. C
3. A
4. D
5. B
6. C
7. D
8. B
9. A
10. C
11. C
12. C
13. C
14. D
15. C
16. B
17. C
18. A
19. D
20. B
21. B
22. C
23. D
24. C
25. C
26. B
27. A
28. C
29. D
30. B
31. B
32. C
33. A
34. D
35. C
36. C
37. C
38. A
39. C
40. C
41. C
42. A
43. A
44. C
45. D
46. C
47. B
48. D
49. C
50. B
51. A
52. C
53. D
54. C
55. D
56. A
57. C
58. B
59. D
60. A
61. B
62. C
63. B
64. D
65. A
66. C
67. D
68. C
69. C
70. C
71. B
72. B
73. C
74. B
75. C
76. D
77. C
78. C
79. A
80. D
81. C
82. A
83. D
84. D
85. C
86. C
87. A
88. C
89. A
90. B
91. D
92. C
93. B
94. D
95. C
96. C

97. D
98. D
99. A
100. D
101. C
102. A
103. A
104. A
105. C
106. C
107. D
108. C
109. D
110. C
111. B
112. C
113. C
114. D
115. D
116. C

Explanations to the Answers for the Sample Test Section 5

Question #1
Answer A is wrong. Technician B is also correct.
Answer B is wrong. Technician A is also correct.
Answer C is correct. Both Technician A and B are correct. Low engine compression will create lower volumetric efficiency. If volumetric efficiency drops, the turbine will slow down due to loss of velocity and exhaust temperature resulting in slower compressor speed and a loss of boost. A wastegate stuck open will by-pass exhaust gases around the turbine and reduce boost.
Answer D is wrong. Both Technician A and B are correct.

Question #2
Answer A is wrong. HC can be measured with a four-gas analyzer.
Answer B is wrong. CO can be measured with a four-gas analyzer.
Answer C is correct. NOx is considered the "fifth" gas.
Answer D is wrong. Oxygen content can be measured with a four-gas analyzer. HC = Unburned fuel. CO = Incomplete combustion; is a good running rich indicator. CO_2 = Engine efficiency complete combustion indicator. O2 = Unused oxygen in the exhaust; is a good lean running indicator. The previous gases are checked using a four-gas analyzer. The fifth gas, NOx, can be checked using a five-gas analyzer.

Question #3
Answer A is correct. Only Technician A is correct. All resistance (carbon core) wires have a specification. Use an ohmmeter to check the wire, disconnected from the spark plug and distributor cap/coil. When checking plug wires, begin with a good visual inspection of the plug wire for damage due to rubbing or heat. Either would cause a secondary voltage leak. Next check the plug wire ends for corrosion and discoloration, which are good indicators of high resistance. Last, check the plug wire resistance value using a digital volt/ohm meter. Compare resistance values with manufacturer specifications. If no specifications are given, Original Equipment Manufacturer wires should be 10,000 ohms or less per foot of length.
Answer B is wrong. Spark plug wires always have some resistance.
Answer C is wrong. Only Technician A is correct.
Answer D is wrong. Only Technician A is correct.

Question #4
Answer A is wrong. Researching service bulletins can save diagnostic time.
Answer B is wrong. Changes to specifications are found is service bulletins.
Answer C is wrong. Service bulletins alert technicians to production changes made during the model year.
Answer D is correct. Year, make, and model identification should already be known to look up service bulletins.

Question #5
Answer A is wrong. Idle speed (throttle plate angle) and ignition timing should always be adjusted prior to final mixture adjustment.
Answer B is correct. Only Technician B is correct. All related engine systems, connections, vacuum hoses, and components should be inspected before any final adjustments are made.
Answer C is wrong. Only Technician B is correct.
Answer D is wrong. Only Technician B is correct.

Question #6
Answer A is wrong. Many radios can lock out if the battery is disconnected and require a code to play again.
Answer B is wrong. Adaptive fuel and idle control values may be erased if the battery is disconnected causing stalling or erratic idle problems.
Answer C is correct. Disconnecting the battery should not prevent air conditioning operation.
Answer D is wrong. Transmission shift adaptive values can be erased when the battery is disconnected, causing a change in shift quality or feel.

Question #7
Answer A is wrong. A flickering test lamp indicates that the module is working properly.
Answer B is wrong. A flickering test lamp indicates the pick-up coil is working properly.
Answer C is wrong. Neither Technician A nor B are correct.
Answer D is correct. Neither Technician A nor B are correct. If the test lamp flickers, primary switching is occurring, indicating the pickup and module are functioning.

Question #8
Answer A is wrong. Some fuel-injection systems provide for air-fuel mixture adjustments.
Answer B is correct. Only Technician B is correct. Some earlier model mechanical and electronic fuel-injection systems have air-fuel mixture adjustment screws.
Answer C is wrong. Only Technician B is correct.
Answer D is wrong. Only Technician B is correct.

Question #9
Answer A is correct. Only Technician A is correct. A vacuum leak will increase the oxygen level in the exhaust stream due to a leaner than normal mixture. HC = Unburned fuel. CO = Incomplete combustion; is a good running rich indicator. CO_2 = Engine efficiency complete combustion indicator. O_2 = Unused oxygen in the exhaust; is a good lean running indicator. A vacuum leak would cause a lean condition showing an increase in O2 readings. Any change to the CO would be a decrease.
Answer B is wrong. A vacuum leak creates a lean mixture and will reduce CO levels in the exhaust.
Answer C is wrong. Only Technician A is correct.
Answer D is wrong. Only Technician A is correct.

Question #10
Answer A is wrong. Sticking valves would not make the vacuum gauge drop.
Answer B is wrong. Advanced ignition timing would not cause vacuum to decrease.
Answer C is correct. An exhaust restriction reduces airflow through the engine and will increase intake manifold pressure and this reduces the vacuum reading.
Answer D is wrong. Rich fuel mixtures would not cause a vacuum decrease off idle. In order for an engine to be able to accept the fresh air/fuel charge, it must be able to expel the inert gases produced during combustion. A restricted exhaust will reduce engine power and maximum speed because of this action. The size of the exhaust restriction will affect the amount of power lost. One of the ways to test for a restricted exhaust is using a vacuum gauge.

Question #11
Answer A is wrong. Technician B is also correct.
Answer B is wrong. Technician A is also correct.
Answer C is correct. Both Technician A and B are correct. The coil has two circuits, a primary and secondary. Both need to be tested for resistance. A dynamic output test using an oscilloscope will show specific voltages.
Answer D is wrong. Both Technician A and B are correct.

Question #12
Answer A is wrong. A proper cooling system inspection involves a pressure test.
Answer B is wrong. A proper cooling system inspection involves inspecting the cooling fan.
Answer C is correct. The evaporator core is not part of the cooling system.
Answer D is wrong. A proper cooling system inspection involves testing the thermostat.

Question #13
Answer A is wrong. Technician B is also correct.
Answer B is wrong. Technician A is also correct.
Answer C is correct. Both Technician A and B are correct. A stethoscope is an excellent pinpoint tool to determining the source of an engine noise. In some instances, the noise needs to be duplicated under certain conditions, such as at a specific temperature or load, or rpm. Noises can be one of the hardest things to diagnose. First, the noise must be pinpointed. In order to pinpoint a noise a stethoscope may be used to amplify the noise. Second, the noise must be identified. The same conditions that produced the noise the first time might need to be met in order to duplicate the noise. A knocking noise, like someone knocking on a door, is usually deep in the engine such as a rod or main bearing. A clicking noise, like someone clicking an ink pen, is usually up top in the valve train. A double knocking noise is usually a wrist pin knock.
Answer D is wrong. Both Technician A and B are correct.

Question #14
Answer A is wrong. When replacing a PROM, always ground yourself to the vehicle, to reduce static discharge.
Answer B is wrong. Being grounded will not erase the PROM.
Answer C is wrong. Neither Technician A nor B are correct.
Answer D is correct. Neither Technician A nor B are correct.

Question #15
Answer A is wrong. Technician B is also correct.
Answer B is wrong. Technician A is also correct.
Answer C is correct. Both Technician A and B are correct. If cylinders located next to each other both indicate a problem, inspect what is common to both. In most cases, it will be a head gasket. If a problem exists in just one cylinder, look for items that control airflow and pressure for that cylinder. When testing compression, remove all spark plugs, install a compression gauge in cylinder number 1, block the throttle open, and crank the engine for four puffs, observing the first puff and last puff. If any readings are low, pour about 1 tablespoon of oil in the low cylinder and recheck compression. If the reading increases, suspect worn piston rings. If the reading does not change, suspect burnt valves or cylinder problems. If two readings adjacent to each other are low, the problem could be a blown head gasket.
Answer D is wrong. Both Technician A and B are correct.

Question #16
Answer A is wrong. The problem may not even be related to the computer control system.
Answer B is correct. The first step in any diagnostic procedure is to verify the customer's concern.
Answer C is wrong. Testing the battery is not a first step.
Answer D is wrong. Bulletins should be consulted after a concern is verified.

Question #17
Answer A is wrong. Technician B is also correct.
Answer B is wrong. Technician A is also correct.
Answer C is correct. Both Technician A and B are correct. Vacuum lines connected to the distributor vacuum advance diaphragm are usually disconnected when setting timing to prevent the distributor from advancing the timing above base and causing the base setting to be retarded. The crankshaft, camshaft, and distributor must be in proper relationship to a specific cylinder, usually #1, when reinstalling a distributor and setting timing.
Answer D is wrong. Both Technician A and B are correct.

Question #18
Answer A is correct. Only Technician A is correct. Most engines will require spark delivery to occur several degrees before the piston reaches top dead-center.
Answer B is wrong. The engine should be timed with the engine on the compression stroke.
Answer C is wrong. Only Technician A is correct.
Answer D is wrong. Only Technician A is correct.

Question #19
Answer A is wrong. TSBs are used to inform technicians of service manual updates.
Answer B is wrong. TSBs are issued for updated parts or service procedures.
Answer C is wrong. TSBs are issued when updated PROM or computer calibrations are released.
Answer D is correct. A service manual, no TSBs should be used to look up torque specifications.

Question #20
Answer A is wrong. A blinking test light connected to the negative side of the coil only confirms primary switching. Low current flow from a voltage drop in the circuit will cause ignition output problems and cannot be diagnosed with a test light. Dwell control and proper triggering while the engine is running are also possible problems that cannot be tested with a test light.
Answer B is correct. Only Technician B is correct. A voltage drop in the primary circuit will reduce secondary circuit performance. A 1-volt drop in the primary can reduce secondary voltage output by as much as 10,000 volts.
Answer C is wrong. Only Technician B is correct.
Answer D is wrong. Only Technician B is correct.

Question #21
Answer A is wrong. Head gaskets can be tested with an emission analyzer.
Answer B is correct. This test is done with labscope or graphing multimeter.
Answer C is wrong. Misfiring cylinders will raise HC emissions.
Answer D is wrong. Emission tests are performed at regular maintenance intervals.

Question #22
Answer A is wrong. Technician B is also correct.
Answer B is wrong. Technician A is also correct.
Answer C is correct. Both Technician A and B are correct. If the timing belt were broken the engine would not run. A jumped tooth on the timing belt would change valve timing and volumetric efficiency.
Answer D is wrong. Both Technician A and B are correct. Correct valve timing is crucial for good drivability and fuel economy. As little as being one tooth out of timing can cause poor fuel economy and poor drivability. Some engines will not run at all just one tooth out of time; however if the timing belt were broken it would definitely be a no start.

Question #23
Answer A is wrong. Black-colored smoke is an indication of oil consumption.
Answer B is wrong. White-colored smoke indicates coolant is being burned; blue is oil.
Answer C is wrong. Neither Technician A nor B are correct.
Answer D is correct. Neither Technician A nor B are correct.

Question #24
Answer A is wrong. Technician B is also correct.
Answer B is wrong. Technician A is also correct.
Answer C is correct. Both Technician A and B are correct. A high-pressure drop during an injector balance test indicates the injector delivered more fuel. The injector could be leaking or may be incorrect for the application.
Answer D is wrong. Both Technician A and B are correct. A fuel injector pressure balance test is a test using a special injector pulser and a pressure gauge. The injector pulser is used to pulse each injector the same amount as the other. By pulsing each injector the same amount and recording the fuel pressure before pulsing the injector we can monitor the amount of pressure drop for each injector. If one injector has a higher pressure drop than the rest, it is running rich probably due to a leaking seat. If one injector has a lower pressure drop than the rest it is running lean, probably due to blockage.

Question #25
Answer A is wrong. Technician B is also correct.
Answer B is wrong. Technician A is also correct.
Answer C is correct. Both Technician A and B are correct. The cooling system may have more than one leak. The cap must be tested to confirm it will hold pressure. There is a specification for the system pressure.
Answer D is wrong. Both Technician A and B are correct.

Question #26
Answer A is wrong. Not all valve adjustments require the engine to be cold. Some manufacturers specify a warm engine.
Answer B is correct. Only Technician B is correct. Many adjustment procedures require the piston at top dead center on the compression stroke as the intake and exhaust valves are closed.
Answer C is wrong. Only Technician B is correct.
Answer D is wrong. Only Technician B is correct.

Question #27
Answer A is correct. Only Technician A is correct. A vacuum leak can set a fuel trim trouble code with no other drivability problems present.
Answer B is wrong. A weak fuel pump would not cause problems at idle when fuel demand is low. Other drivability problems like lack of power would be present.
Answer C is wrong. Only Technician A is correct.
Answer D is wrong. Only Technician A is correct.

Question #28
Answer A is wrong. This detects pinging and is used to retard timing.
Answer B is wrong. Engine speed is used to calculate advance curve.
Answer C is correct. Power steering load input is used for idle compensation, not timing.
Answer D is wrong. Engine load is used to calculate timing advance.

Question #29
Answer A is wrong. The fuel pump is not part of this test.
Answer B is wrong. Grounding the wire would result in a direct short.
Answer C is wrong. Grounding the wire would result in a direct short.
Answer D is correct. An approved spark tester will require the coil to put out approximately 25 kV, usually sufficient for all engines.

Question #30
Answer A is wrong. An air manifold restriction will not cause a backfire on deceleration.
Answer B is correct. A malfunctioning diverter valve will cause a backfire.
Answer C is wrong. An exhaust manifold check valve will not cause a backfire on deceleration.
Answer D is wrong. An output pressure air pump will not cause a backfire on deceleration.

Question #31
Answer B is correct. Timing is common to all cylinders.
Answer A is wrong. Defective spark plugs can be diagnosed with a cylinder balance test.
Answer C is wrong. Faulty ignition wires can be determined with cylinder power balance tests.
Answer D is wrong. Burned valves can be diagnosed with cylinder power balance tests.

Question #32
Answer A is wrong Technician B is also correct.
Answer B is wrong. Technician A is also correct.
Answer C is correct. Both Technician A and B are correct. A weak coil may provide enough spark to fire normally at an idle, but not enough to meet demands at cruise or when accelerating. If the problem existed long enough, there may be a DTC stored.
Answer D is wrong. Both Technician A and B are correct.

Question #33
Answer A is correct. Hotter temperatures result in more resistance, thus causing the fan to move more air across the radiator.
Answer B is wrong. It has less resistance cold, when fan-blown air is not needed, to improve performance and fuel economy.
Answer C is wrong. It should have movement.
Answer D is wrong. The clutch should have resistance.

Question #34
Answer A is wrong. If the pickup coil leads are moved, there should be no erratic reading on the ohmmeter. If there is, look for damaged pickup coil wires.
Answer B is wrong. An infinite reading indicates an open circuit.
Answer C is wrong. Neither Technician A nor B are correct. When testing a pickup coil, the first step is to measure the pickup coil resistance. Pickup resistance values range from 150 ohms to 1,500 ohms depending on the manufacturer. If the distributor that houses the pickup coil has a vacuum advance, be sure to apply vacuum while watching the resistance reading throughout the entire range of operation; any erratic readings are bad.
Answer D is correct. Neither Technician A nor B are correct.

Question #35
Answer A is wrong. This is a lean condition.
Answer B is wrong. A DTC may be set.
Answer C is correct. This is not a rich condition.
Answer D is wrong. The O2 sensor is operating properly. An O2 sensor can fail in several ways. It can fail producing no voltage, it can fail in the lean biased condition, it can fail in the rich biased condition, or it can just become lazy. On a lean biased condition the sensor operates, it just stays under 500 mV more than it is above 500 mV. On a rich biased condition the sensor is above 500 mV more than it is under 500 mV. When an O2 sensor is lazy it switches back and forth from rich to lean, it just switches very slowly.

Question #36
Answer A is wrong. Technician B is also correct.
Answer B is wrong. Technician A is also correct.
Answer C is correct. Both Technician A and B are correct. Any vacuum leak will change the air-fuel mixture, resulting in loss of performance and fuel economy. Controlled use of propane gas can help confirm and locate the source of vacuum leaks.
Answer D is wrong. Both Technician A and B are correct.

Question #37
Answer A is wrong. Technician B is also correct.
Answer B is wrong. Technician A is also correct.
Answer C is correct. Both Technician A and B are correct. Not installing dielectric compound on an ignition module can cause repeat failure from excessive heat build-up in the module. A shorted ignition coil primary winding will increase current flow through the module and cause repeat transistor failure. Some ignition modules require the use of dielectric grease on the bottom of the module to assist in heat transfer. The power transistor gets very hot during operation and without the dielectric grease the module would overheat and fail. When replacing a failed module always check the ignition coil primary windings for a short, otherwise the new module may fail due to high current draws from the primary circuit.
Answer D is wrong. Both Technician A and B are correct.

Question #38
Answer A is correct. Only Technician A is correct. Shutting off fuel injectors at high rpm can protect the engine from over-revving or limit maximum vehicle speed.
Answer B is wrong. Most fuel-injection systems turn off fuel injectors under certain conditions such as decel to reduce fuel consumption and emissions and protect the catalytic converter from over-rich mixtures.
Answer C is wrong. Only Technician A is correct.
Answer D is wrong. Only Technician A is correct.

Question #39

Answer A is wrong. Technician B is also correct.

Answer B is wrong. Technician A is also correct.

Answer C is correct. Both Technician A and B are correct. Depressing the throttle fully during cranking will make the computer enter clear flood mode and lean out the air/fuel mixture. Low-system voltage will enable battery voltage correction programming in the computer that will increase injector pulse width commands due to slower injector opening from low-system voltage. On many automobiles there is an operating strategy called clear flood. Clear flood is to assist the operator in starting a flooded vehicle. For clear flood to be initiated the operator holds the accelerator at WOT while cranking the engine. The PCM lowers the injector pulse width or shuts off the injectors altogether depending on manufacturer. This allows the engine to start without any additional fuel being injected into the engine. Some manufacturers increase the injector pulse width when system voltage is low to counteract slow injector operation due to low voltage.

Answer D is wrong. Both Technician A and B are correct.

Question #40

Answer A is wrong. A throttle position sensor problem may cause a hesitation.

Answer B is wrong. A restricted MAP sensor hose may cause a hesitation problem.

Answer C is correct. A disconnected fuel-pressure regulator vacuum line will raise fuel pressure and enrich the air-fuel mixture. This would not cause a hesitation.

Answer D is wrong. A cracked air intake hose will allow unmetered air into the engine during acceleration and cause a hesitation.

Question #41

Answer A is wrong. Technician B is also correct.

Answer B is wrong. Technician A is also correct.

Answer C is correct. Both Technician A and B are correct. If the plunger is stuck open, it will be an equivalent of a vacuum leak, and will draw excessive crankcase vapors and possibly engine oil into the intake.

Answer D is wrong. Both Technician A and B are correct.

Question #42

Answer A is correct. Only Technician A is correct. Tight bushings can cause added drag on the starter. A starter free-running test is used to test starter ability without any outside load being placed on the starter. The starter is engaged and checked for current draw and rpm. This test is often performed after a starter is rebuilt. Worn brushes will typically cause a no engagement complaint because the brushes make contact intermittently. Tight brushes, however, can cause the rpm to be lower than specifications and cause the current draw to be elevated due to the dragging starter.

Answer B is wrong. Worn brushes will not cause the starter to have low rpms.

Answer C is wrong. Only Technician A is correct.

Answer D is wrong. Only Technician A is correct.

Question #43

Answer A is correct. Fuel tank leaks caused by defective tank straps are rare.

Answer B is wrong. Road damage often causes fuel tank damage.

Answer C is wrong. Defective seams can be a cause of fuel leaks.

Answer D is wrong. Corrosion, after it is through the tank, will cause a fuel leak.

Question #44

Answer A is wrong. Nylon hoses should always be inspected for kinks.

Answer B is wrong. Loose fittings should be inspected for fuel leakage.

Answer C is correct. Discoloration will not affect performance of the component.

Answer D is wrong. Nylon can be scratched easily and should be inspected.

Question #45
Answer A is wrong. Many TBI systems do not provide a Schrader test port for fuel system testing.
Answer B is wrong. A plugged filter cannot cause high fuel pressure because the pump is behind the filter and because the pressure regulator controls fuel pressure.
Answer C is wrong. Neither Technician A nor B are correct.
Answer D is correct. Neither Technician A nor B are correct.

Question #46
Answer A is wrong. Technician B is also correct.
Answer B is wrong. Technician A is also correct.
Answer C is correct. Both Technicians A and B are correct. The pressure test is done to prove pump operation, thus confirming power, ground, and pump action. Specifications are published for fuel pressure and volume. Checking fuel pump pressure is part of diagnosing the fuel system. However, this is only part of the diagnosis. Fuel pump pressure, volume, and even RPM are all part of a complete fuel system diagnosis. It's very possible to have good pressure but not enough volume to properly run the engine.
Answer D is wrong. Both Technician A and B are correct.

Question #47
Answer A is wrong. A head gasket leak would cause air leakage to an adjacent cylinder or the cooling system, not the crankcase.
Answer B is correct. Only Technician B is correct. Damaged piston rings would cause air leakage into the crankcase and this would escape through the oil filler cap or PCV opening in the valve cover.
Answer C is wrong. Only Technician B is correct.
Answer D is wrong. Only Technician B is correct.

Question #48
Answer A is wrong. High fuel pressure is not the LEAST-Likely cause of poor fuel mileage.
Answer B is wrong. A disconnected regulator is not the LEAST-Likely cause of poor fuel mileage.
Answer C is wrong. A partially plugged exhaust is not the LEAST-Likely cause of poor fuel mileage.
Answer D is correct. A malfunctioning diverter valve will cause a backfire.

Question #49
Answer A is wrong. Technician B is also correct.
Answer B is wrong. Technician A is also correct.
Answer C is correct. Both Technician A and B are correct. Deposits can build up in passageways in the throttle body, thus requiring removal. Any build up on the body or plates will result in restricted or disturbed airflow, causing performance problems. The buildup of carbon on the throttle body is a common problem on a multiport fuel-injected system. Effects of a dirty throttle body are usually a vehicle that idles rough or stalls at idle. Some throttle bodies can be easily cleaned on the vehicle. Some throttle bodies are difficult to clean without removal. A dirty throttle body can be diagnosed by sight or by using a scan tool and looking at the IAC steps. Steps over 20 should be followed up with a visual throttle body inspection.
Answer D is wrong. Both Technician A and B are correct.

Question #50
Answer A is wrong. A fuel-pressure test will not confirm that the fuel pump can deliver adequate volume. A volume test must also be performed.
Answer B is correct. Only Technician B is correct. It is possible for a fuel injector to have proper electrical operation but not deliver fuel properly; such as a restriction or bad spray pattern.
Answer C is wrong. Only Technician B is correct.
Answer D is wrong. Only Technician B is correct.

Question #51
Answer A is correct. Combustion beginning before the timed spark is called preignition.
Answer B is wrong. Over-advanced timing should not start combustion.
Answer C is wrong. Dieseling is the after run that occurs in a gasoline engine when air and fuel are ignited in the combustion chamber by hot spots after the engine is turned off.
Answer D is wrong. Lean burn combustion is a technology, not an abnormal combustion event.

Question #52
Answer A is wrong. Answer B is also correct.
Answer B is wrong. Answer A is also correct.
Answer C is correct. Both Technician A and B are correct. Either condition will cause a rich mixture.
Answer D is wrong. Both Technician A and B are correct.

Question #53
Answer A is wrong. A MAP sensor is directly related to the air-fuel mixture.
Answer B is wrong. A defective MAP sensor can cause surging of the engine.
Answer C is wrong. If the MAP sensor is bad, it can cause a rich air-fuel mixture.
Answer D is correct. The MAP sensor is the load sensing sensor of a vehicle using the speed density method of load calculation. The MAP sensor is continually monitoring the manifold absolute pressure and sending these varying changes to the PCM to use in pulse with calculations. A faulty MAP sensor can cause problems such as a no start, poor fuel economy, engine surging, and varying air-fuel ratios. A MAP sensor is not likely to cause any excessive idle speed concerns.

Question #54
Answer A is wrong. Checking the spray pattern is not the easiest test to perform.
Answer B is wrong. There is only one cold start injector; therefore, it cannot be compared to another.
Answer C is correct. This, in most cases, is the easiest test to do.
Answer D is wrong. It is easier to perform a resistance value check.

Question #55
Answer A is wrong. A straightedge should be used with feeler gauges.
Answer B is wrong. A straightedge should be used with feeler gauges.
Answer C is wrong. Neither Technician A nor B are correct.
Answer D is correct. Neither Technician A nor B are correct.

Question #56
Answer A is correct. Only Technician A is correct. Introducing EGR gases into the engine at idle will cause severe idle roughness or stall the engine. Opening the EGR valve at idle allows inert gas to be routed back into the intake manifold. This will cause the engine to run rough or even stall at idle. Normal operation of the EGR valve would be on a warmed up engine during cruising speeds. The best RPM to test EGR operation would be at idle where it would make the change in rpm.
Answer B is wrong. It is best to determine if EGR gases are reaching the combustion chamber by opening the EGR valve during engine idle because higher rpm conditions might not show a change when the EGR valve is opened.
Answer C is wrong. Only Technician A is correct.
Answer D is wrong. Only Technician A is correct.

Question #57
Answer A is wrong. A defective EVVRV vacuum regulator could prevent the switch from closing and cause a constant infinite resistance reading.
Answer B is wrong. A defective EGR valve could leak vacuum and prevent the diagnostic switch from closing. This could cause the problem mentioned.
Answer C is correct. A switch that remains constantly open does not indicate normal operation.
Answer D is wrong. A broken or disconnected vacuum line will prevent the switch from closing.

Question #58

Answer A is wrong. A higher resistance value on a negative temperature coefficient (NTC) thermistor coolant sensor means a cold engine signal, not a warmer engine input.

Answer B is correct. Only Technician B is correct. Higher resistance in the coolant sensor means a colder then actual engine temperature. The computer will deliver a richer air/fuel mixture than needed and hard, warm engine starting will result.

Answer C is wrong. Only Technician B is correct.

Answer D is wrong. Only Technician B is correct.

Question #59

Answer A is wrong. Air pump output constantly directed to the exhaust manifold will Most-Likely decrease fuel economy because the computer will see a lean exhaust signal and increase fuel delivery.

Answer B is wrong. Air injected into the exhaust manifold would not cause pinging on acceleration.

Answer C is wrong. Air injected into the exhaust manifold constantly would not cause engine overheating. The function of the oxygen sensor is to measure the amount of oxygen in the exhaust stream. In the event that secondary air is misrouted upstream to the exhaust manifold, the oxygen sensor will see an excess of oxygen. This will cause the oxygen sensor to send a constant lean signal to the PCM, in turn causing a rich command to the injectors.

Answer D is correct. Air injected into the exhaust manifold constantly will cause the oxygen sensor to send a continuous lean output or low voltage signal.

Question #60

Answer A is correct. Only Technician A is correct. A dual bed catalytic converter often uses secondary air injection as an additional source of oxygen to oxidize HC and CO in the rear bed of the converter.

Answer B is wrong. Many manufacturers use secondary air injection for the catalytic converter.

Answer C is wrong. Only Technician A is correct.

Answer D is wrong. Only Technician A is correct.

Question #61

Answer A is wrong. The evaporative emission system recovers HC emissions from the fuel tank.

Answer B is correct. Only Technician B is correct. The canister traps vapor and purges while driving. The vapors are carefully metered and are eliminated through the combustion process. The function of the evaporative-emission systems is to prevent the release of hydrocarbons into the atmosphere. It traps and holds gasoline vapors. The EVAP system consist of a vapor canister, purge valve, pressure control valve, rollover check valve, fuel tank, and cap. Typical EVAP-related engine performance problems are poor fuel economy and poor performance.

Answer C is wrong. Only Technician B is correct.

Answer D is wrong. Only Technician B is correct.

Question #62

Answer A is wrong. Technician B is also correct.

Answer B is wrong. Technician A is also correct.

Answer C is correct. Both Technician A and B are correct. Some canisters have a replaceable filter. A restricted canister filter can prevent fresh airflow through the canister, causing ineffective purging. It should be checked or replaced at recommended service intervals. Gas caps can be checked for pressure and vacuum holds. There is equipment and specifications available for this test.

Answer D is wrong. Both Technician A and B are correct.

Question #63

Answer A is wrong. An open VREF wire to a coolant sensor will set a DTC.

Answer B is correct. Operating the vehicle in extremely cold weather will not set a DTC.

Answer C is wrong. A shorted VREF wire to a coolant sensor will set a DTC.

Answer D is wrong. An out of range voltage input will set a DTC.

Question #64
Answer A is wrong. Improper cooling affects the life of the bearings.
Answer B is wrong. Not changing the oil frequently causes sludge buildup.
Answer C is wrong. A dirty air cleaner will contaminate the air intake with dust and dirt, which shorten the life of the turbocharger. The life of a turbo depends on many factors. Keeping the bearings from overheating is crucial. Heat is dissipated through the coolant as well as the lubricating oil. Oil changes at scheduled intervals are extremely important for turbo life. And finally, filtration of incoming air is needed to prevent compressor abrasion due to airborne contaminates.
Answer D is correct. Exhaust leaks do not cause turbo damage.

Question #65
Answer A is correct. Only Technician A is correct. Incorrect inputs can result in performance or emission problems. The throttle position sensor is an input to the PCM telling throttle position and speed of throttle opening. In the event the TPS fails, it might set one of many diagnostic trouble codes. However, it might not set any. If the sensor gets a bad spot in the potentiometer it could cause a hesitation without setting a code. The use of a DVOM or labscope would show such a problem.
Answer B is wrong. A faulty TPS sensor does not always cause a fault code.
Answer C is wrong. Only Technician A is correct.
Answer D is wrong. Only Technician A is correct.

Question #66
Answer A is wrong. Technician B is also correct.
Answer B is wrong. Technician A is also correct.
Answer C is correct. Both Technician A and B are correct. Exhaust gases that are restricted will affect volumetric efficiency, interfering with the new incoming air/fuel charge. In order for an engine to be able to accept the fresh air/fuel charge, it must be able to expel the inert gases produced during combustion. A restricted exhaust will reduce engine power and maximum speed because of this action. The size of the exhaust restriction will affect the amount of power lost.
Answer D is wrong. Both Technician A and B are correct.

Question #67
Answer A is wrong. Since the valve is located downstream from the throttle plates, there will always be vacuum at the PCV valve.
Answer B is wrong. There should be a noise when the valve is shaken, proving the plunger is not stuck.
Answer C is wrong. Neither Technician A nor B are correct.
Answer D is correct. Neither Technician A nor B are correct.

Question #68
Answer A is wrong. Technician B is also correct.
Answer B is wrong. Technician A is also correct.
Answer C is correct. Both Technician A and B are correct. A vacuum leak or bad oxygen sensor can cause abnormal fuel trim corrections at idle.
Answer D is wrong. Both Technician A and B are correct.

Question #69
Answer A is wrong. Technician B is also correct.
Answer B is wrong. Technician A is also correct.
Answer C is correct. Both Technician A and B are correct. A restricted converter will limit engine power and top speed because it reduces airflow or volumetric efficiency of the engine. Rapping on the converter shell with a rubber mallet can identify a broken monolith in a monolithic converter.
Answer D is wrong. Both Technician A and B are correct.

Question #70
Answer A is wrong. Technician B is also correct.
Answer B is wrong. Technician A is also correct.
Answer C is correct. Both Technician A and B are correct. A MAP sensor should hold vacuum while testing. MAP sensors can be either analog or digital output sensors. Knowing which type is used is necessary so that the correct test equipment and procedures are used. The MAP sensor tells the PCM the load on the engine. Under heavy load, vacuum will be low, while during a decal the vacuum will be high. When testing a MAP sensor with a vacuum gauge and voltmeter, the MAP sensor should hold vacuum. Most MAP sensors produce an analog signal; however, some vehicles use a MAP sensor that produces a frequency and a digital signal.
Answer D is wrong. Both Technician A and B are correct.

Question #71
Answer A is wrong. Accurate testing of the oxygen sensor cannot be performed with the sensor removed.
Answer B is correct. Only Technician B is correct. A scan tool is useful for checking for the presence of trouble codes as well as monitoring oxygen sensor voltage. One of the best ways to check the operation of an O2 sensor is to use a scan tool to check for codes then check scan tool data stream. While looking at O2 sensor voltage with a scan tool slowly restrict the vehicle's airflow, causing the engine to go rich. The O2 sensor voltage should increase on an operating sensor. Then cause a vacuum leak by slowly disconnecting a large vacuum hose. The O2 sensor voltage should drop to close to zero volts.
Answer C is wrong. Only Technician B is correct.
Answer D is wrong. Only Technician B is correct.

Question #72
Answer A is wrong. Damaged carbon ignition wires cause high resistance.
Answer B is correct. The ignition module is common to all cylinders and is part of the primary ignition circuit.
Answer C is wrong. Corroded spark plug wire ends will cause high resistance.
The scope pattern shown is a parade pattern that shows all the cylinders side by side. Only one cylinder has high resistance, which is seen by the high firing line. Bad spark plug wires are very common causes of high resistance in the secondary pattern. Spark plugs can fail due to interference with other components, internal breakdown, and corrosion. Another cause of high firing voltage is lean air/fuel mixtures and excessive spark plug gap.

Question #73
Answer A is wrong. Technician B is also correct.
Answer B is wrong. Technician A is also correct.
Answer C is correct. Both Technician A and B are correct. A voltage drop test is a more accurate way of finding resistance in a circuit because you test the circuit at work. Voltage drop is defined as the voltage used pushing current through a resistance. A battery cable should pose very little resistance to current flow so performing a voltage drop test across the cable determines if excessive resistance is present in the cable. A voltage drop greater than .5 volts across a positive battery cable is excessive.
Answer D is wrong. Both Technician A and B are correct.

Question #74
Answer A is wrong. The first step in a diagnostic procedure should be to verify the customer concern. Only if a concern is related to a system that has self-diagnostic capabilities does it make sense to check for trouble codes.
Answer B is correct. Only Technician B is correct. If a customer complaint cannot be verified, then diagnostic testing may not produce any results.
Answer C is wrong. Only Technician B is correct.
Answer D is wrong. Only Technician B is correct.

Question #75
Answer A is wrong. Technician B is also correct.
Answer B is wrong. Technician A is also correct.
Answer C is correct. Both Technician A and B are correct. A visual inspection of all vacuum hose and electrical connections in addition to checking for related DTCs are essential diagnostic steps.
Answer D is wrong. Both Technician A and B are correct.

Question #76
Answer A is wrong. A rich fuel mixture would not cause knocking.
Answer B is wrong. A bad crank sensor would not cause knocking.
Answer C is wrong. High manifold vacuum would not cause knocking. Spark knock is generally the result of overheating of the combustion chamber. Many things can cause overheating of the combustion chamber. A lean air/fuel mixture burns much hotter than a rich mixture and can cause spark knock. The EGR system's job is to lower combustion chamber temperature by allowing the recirculation of inert gases back into the combustion chamber. Most vehicles will spark knock if the EGR valve is disabled.
Answer D is correct. Restricted EGR would increase combustion temperatures, causing knock and NOx.

Question #77
Answer A is wrong. Technician B is also correct.
Answer B is wrong. Technician A is also correct.
Answer C is correct. Both Technician A and B are correct. Enabling criteria are the specific operating parameters that must be met before a monitor will run. Many EVAP monitors for instance require a cold start with ambient temperature below a certain value. Pending conditions are any circumstances that may prevent a monitor from running properly, such as an oxygen sensor fault not allowing a catalyst monitor from running.
Answer D is wrong. Both Technician A and B are correct.

Question #78
Answer A is wrong. Technician B is also correct.
Answer B is wrong. Technician A is also correct.
Answer C is correct. Both Technician A and B are correct. As components wear, clearance is often increased beyond an allowable tolerance, thus creating a noise. It is important to determine the exact source of the noise before any disassembly, due to the noise "traveling," making it seem like it is coming from another area.
Answer D is wrong. Both Technician A and B are correct.

Question #79
Answer A is correct. Only Technician A is correct. Retarded ignition timing will cause peak combustion pressure to occur too late in the power stroke. This reduces crankshaft velocity and rpm and causes low-intake manifold vacuum.
Answer B is wrong. Connecting the vacuum gauge to a ported vacuum source will read vacuum off idle when the throttle is opened, not at idle.
Answer C is wrong. Only Technician A is correct.
Answer D is wrong. Only Technician A is correct.

Question #80
Answer A is wrong. Worn valves will not properly seal the combustion chamber and cause a compression loss.
Answer B is wrong. Worn piston rings are a common cause of low cylinder compression.
Answer C is wrong. A blown head gasket can cause a loss of cylinder compression. The cause of low cylinder compression is due to escaping pressure from the combustion chamber. Worn rings are a typical cause of low compression. The rings are what seal the piston in the cylinder. The head gasket, which seals the head to the block, can develop a leak at the combustion chamber (blown head gasket) allowing the escape of combustion pressure. A burnt valve will also cause the escape of compression gases. If the valve guilds were worn they would still seal the combustion chamber when the valves are shut.
Answer D is correct. Worn valve guides often cause excessive oil consumption but may not affect valve sealing therefore they often will not cause lower compression.

Question #81
Answer A is wrong. Technician B is also correct.
Answer B is wrong. Technician A is also correct.
Answer C is correct. Both Technician A and B are correct. Any change in the delivery of the air/fuel charge will result in a performance or economy complaint.
Answer D is wrong. Both Technician A and B are correct.

Question #82
Answer A is correct. Only Technician A is correct. On vehicles with adjustable timing, specifications and timing setup procedures are on the under hood label.
Answer B is wrong. Timing lights must be connected to only one specified plug wire (usually number one).
Answer C is wrong. Only Technician A is correct.
Answer D is wrong. Only Technician A is correct.

Question #83
Answer A is wrong. A kink in a fuel line can cause reduced fuel flow and aeration that may cause drivability problems.
Answer B is wrong. Nylon fuel line can be spliced and end connectors replaced using approved repair kits.
Answer C is wrong. Neither Technician A nor B are correct.
Answer D is correct. Neither Technician A nor B is correct.

Question #84
Answer A is wrong. Resistance is an approved method for checking a coolant sensor.
Answer B is wrong. Checking the return voltage is an approved method for checking a coolant sensor.
Answer C is wrong. The coolant temperature sensor can be placed in heated water while measuring the resistance. When a vehicle comes in with a coolant sensor system fault in the PCM, one of many things is checked depending on the diagnostic flow chart being used. Some flow charts have us check the resistance of the sensor compared to engine temperature. Always compare the ECT value with the IAT value after the vehicle has had time for all the engine temperatures to equalize. On some diagnoses it may be necessary to submerge the sensor in heated water and watch for a decrease in resistance (negative temperature coefficient).
Answer D is correct. Diode check is not part of testing the coolant sensor/or circuitry.

Question #85
Answer A is wrong. Technician B is also correct.
Answer B is wrong. Technician A is also correct.
Answer C is correct. Both Technician A and B are correct. The canister filter is a normal maintenance item. The filler cap must be checked for vacuum and pressure holds.
Answer D is wrong. Both Technician A and B are correct.

Question #86
Answer A is wrong. Technician B is also correct.
Answer B is wrong. Technician A is also correct.
Answer C is correct. Both Technician A and B are correct. A leak in the diaphragm will cause the fuel pump to not be able to draw the proper volume of fuel, or cause it to draw air. It must be able to build pressure, and maintain pressure. There are three checks made on a mechanical fuel pump using a vacuum/pressure gauge. First, on the inlet side, we check pump vacuum. This tells us the pump's ability to draw fuel, checking the diaphragm and the inlet and outlet check valves. Second, we check for adequate pressure, usually 4 to 8 psi. Third, we check capacity (flow), which should be at least 2 pints per minute.
Answer D is wrong. Both Technician A and B are correct.

Question #87
Answer A is correct. It is best to check the fuel pressure and volume first.
Answer B is wrong. There are no fuel pump diagnostic codes.
Answer C is wrong. It is easier to check the fuel pressure and volume first.
Answer D is wrong. Inspecting the fuel lines should only be done if they are suspect. Anytime a faulty fuel pump is suspected the technician should begin by checking the fuel pump's pressure and volume. It's possible for a fuel pump to have good pressure and be low enough in volume to cause a drivability problem.

Question #88
Answer A is wrong. Technician B is also correct.
Answer B is wrong. Technician A is also correct.
Answer C is correct. Both Technician A and B are correct. The cold start injector is operated through a thermo-time circuit.
Answer D is wrong. Both Technician A and B are correct.

Question #89
Answer A is correct.
Answer B is wrong. A dirty air filter will not cause a no-start condition.
Answer C is wrong. Poor fuel economy would be noticed before engine wear.
Answer D is wrong. Excessive engine wear would have to occur before the engine would use an excessive amount of oil. An air filter is needed to filter out air-borne abrasives and contaminants and prevent them from entering the engine. As the filter gets dirty the flow of air becomes restricted. This will cause a loss of power and a loss in fuel economy. To check the condition on an air filter use a shop light to shine through one side of the filter then check for light intensity on the other side. The light will show just how dirty the filter is.

Question #90
Answer A is wrong. Any voltage drop reading of the starter control circuit above .5 volts is considered excessive.
Answer B is correct. Only Technician B is correct. If voltage drop in a starter control circuit is excessive, then individual components such as the ignition switch or starter relay, and their related wiring will need to be tested to isolate the exact fault.
Answer C is wrong. Only Technician B is correct.
Answer D is wrong. Only Technician B is correct.

Question #91
Answer A is wrong. A damage turbo bearing will limit the impeller speed.
Answer B is wrong. An open wastegate will bypass exhaust around the impeller and reduce its speed.
Answer C is wrong. A leak limits maximum pressure in the cylinders.
Answer D is correct. Low coolant temperature would not have a measurable effect on turbo boost output.

Question #92
Answer A is wrong. Technician B is also correct.
Answer B is wrong. Technician A is also correct.
Answer C is correct Both Technician A and B are correct. The alternator connectors should be inspected any time an alternator is being checked or replaced. Some replacement alternators come with a new connector.
Answer D is wrong. Both Technician A and B are correct.

Question #93
Answer A is wrong. An open PCV valve will lean the mixture, not richen it.
Answer B is correct. Only Technician B is correct. Crankcase pressure builds, causing blowby.
Answer C is wrong. Only Technician B is correct.
Answer D is wrong. Only Technician B is correct.

Question #94
Answer A is wrong. Thirty seconds is too long. Depending on the manufacturer, fifteen seconds is usually the maximum.
Answer B is wrong. Never test a battery at the cold crank rating. The specification is ½ the CCA.
Answer C is wrong. Neither Technician A nor B are correct.
Answer D is correct. Neither Technician A nor B are correct. When doing a battery capacity test, load the battery until the ammeter reads one-half the cold cranking rating or three times the ampere-hour rating. Maintain this load position for 15 seconds. If the battery voltage is less than the minimum load voltage (9.6 volts at 70 degrees and above) at the end of the load test, the battery should be charged and the load test repeated. After charging and a repeat load test, if the battery voltage is less than the minimum load voltage, the battery should be replaced.

Question #95
Answer A is wrong. Technician B is also correct.
Answer B is wrong. Technician A is also correct.
Answer C is correct. Both Technician A and B are correct. A stuck open EGR valve causes diluted air-fuel mixture, while an EGR system that does not operate causes high combustion chamber temperatures, resulting in detonation and NOx output. Opening the EGR valve at idle allows inert gas to be routed back into the intake manifold. This will cause the engine to run rough or even stall at idle. Normal operation of the EGR valve would be on a warmed up engine during cruising speeds. The best rpm to test EGR operation would be at idle where it would make the change in rpm. The EGR valve is used to lower NOX emissions by lowering combustion chamber temperatures. Some vehicles will even produce a detonation problem due to combustion chamber over heating.
Answer D is wrong. Both Technician A and B are correct.

Question #96
Answer A is wrong. Cold-cranking amps is a battery rating.
Answer B is wrong. Ampere hours is a battery rating.
Answer C is correct. Reserve amps is not a battery rating.
Answer D is wrong. Reserve minutes is a battery rating.

Question #97
Answer A is wrong. A stuck closed EGR valve can cause spark knock.
Answer B is wrong. Poor fuel quality can cause spark knock.
Answer C is wrong. Carbon build-up in the combustion chamber can cause spark knock.
Answer D is correct. A stuck open EGR valve would not cause spark knock. Other drivability problems would be present.

Question #98
Answer A is wrong. An infinite reading indicates only an open winding.
Answer B is wrong. An infinite reading does not indicate a shorted winding.
Answer C is wrong. Neither Technician A nor B are correct.
Answer D is correct. Neither Technician A nor B are correct.

Question #99
Answer A is correct. Only Technician A is correct. A change in vacuum will either open or close the valve for tank vapor control to the canister.
Answer B is wrong. The valve does not have to be installed on the vehicle.
Answer C is wrong. Only Technician A is correct.
Answer D is wrong. Only Technician A is correct.

Question #100

Answer A is wrong. Diagnostic trouble codes do not necessarily indicate which component to replace but point to the circuit or system at fault. Pinpoint testing is necessary to determine the exact problem.

Answer B is wrong. The emissions reminder light is preset to illuminate at a certain mileage or elapsed time, not when a fault occurs. The malfunction indicator light (MIL), or check engine light, illuminates when a fault occurs in the emission control system.

Answer C is wrong. Neither Technician A nor B are correct. A diagnostic trouble code only tells you what system or circuit the failure occurred in. After the DTC is retrieved, the appropriate diagnostic flow chart should be obtained and followed to properly diagnose the failure. The emission reminder lamp is used by some manufacturers on some vehicles to notify the driver that some emission-related service is required. These lamps are tripped by mileage.

Answer D is correct. Neither Technician A nor B are correct.

Question #101

Answer A is wrong. Technician B is also correct.

Answer B is wrong. Technician A is also correct.

Answer C is correct. Both Technician A and B are correct. A restricted MAF inlet screen will cause an improper response from the sensor and a hesitation under acceleration. A TPS signal dropout will cause the computer to reduce fuel flow thinking the throttle was closing, and this can cause a hesitation under acceleration.

Answer D is wrong. Both Technician A and B are correct.

Question #102

Answer A is correct. Only Technician A is correct. The pulse width increases to richen the mixture.

Answer B is wrong. Short-term fuel is subtracted or reduced as a result of a rich condition.

Answer C is wrong. Only Technician A is correct.

Answer D is wrong. Only Technician A is correct.

Question #103

Answer A is correct. Only Technician A is correct. A digital voltmeter (high impedance) can be used to monitor O_2 sensor voltage.

Answer B is wrong. An O2 sensor cannot be checked with diode tester or an ohmmeter.

Answer C is wrong. Only Technician A is correct.

Answer D is wrong. Only Technician A is correct.

Question #104

Answer A is correct. Only Technician A is correct. Use only a digital, high impedance voltmeter.

Answer B is wrong. A test light cannot test an O2 sensor, nor should you try.

Answer C is wrong. Only Technician A is correct.

Answer D is wrong. Only Technician A is correct.

Question #105

Answer A is wrong. Technician B is also correct.

Answer B is wrong. Technician A is also correct.

Answer C is correct. Both Technician A and B are correct. Glazed, cracked, or deteriorated belts can make noise. The proper fix is to replace the belt. In any case, the proper belt tension adjustment must be made.

Answer D is wrong. Both Technician A and B are correct.

Question #106

Answer A is wrong. Corrosion on connector terminals will increase resistance and voltage drop.

Answer B is wrong. Cut strands on multiple strand wire reduce the wires cross-section and increases resistance and voltage drop.

Answer C is correct. Using heavier gauge wire increases current capacity and lowers resistance.

Answer D is wrong. Spread female wire terminals will increase resistance and voltage drop. As current passes through a resistance, heat is generated. With this generation of heat comes a loss in voltage. The voltage that is converted to heat by a resistance is called the voltage drop. Corrosion on connector terminals, cut strands in a multiple strand wire, and spread female terminals in a connector are all sources of voltage drop.

Question #107

Answer A is wrong. Static electricity will harm a PCM, regardless of whether or not the battery cables are connected.

Answer B is wrong. Proper handling procedures for static sensitive components calls for wearing a grounding strap on your wrist connected to a body ground source whenever handling these components.

Answer C is wrong. Neither Technician is correct.

Answer D is correct. Neither Technician A nor B are correct. When working with the PCM static electricity is always a concern. When replacing a PCM never touch the thermals of the PCM. It's good practice to use a special grounding strap to ground yourself to the vehicle before removing the PCM. If not, always touch a metal part of the vehicle to discharge any static charge that may be in your body before removing the PCM.

Question #108

Answer A is wrong. Technician B is also correct.

Answer B is wrong. Technician A is also correct.

Answer C is correct. Both Technician A and B are correct. For example, the body controller may use data from the PCM to determine when to turn dashboard warning lights on, i.e., temperature. The transmission or climate control systems may also be tied to the powertrain control module.

Answer D is wrong. Both Technician A and B are correct.

Question #109

Answer A is wrong. The ignition module does not need an rpm signal.

Answer B is wrong. There is no such thing as Hall effect timing.

Answer C is wrong. The ignition module does not discriminate between cylinders.

Answer D is correct. The signal from the PCM is the result of computed inputs for proper timing advance. This is the signal the ignition module uses to fire the coil at the proper time.

Question #110

Answer A is wrong. Technician B is also right.

Answer B is wrong. Technician A is also right.

Answer C is correct. Both Technician A and B are correct. While the preferred method of clearing diagnostic trouble codes is to use a scan tool, some manufacturer's procedures say to disconnect the PCM fuse.

Answer D is wrong. Both Technician A and B are correct.

Question #111

Answer A is wrong. It is good practice to perform a voltage drop check before replacing a component.

Answer B is correct. Only Technician B is correct. It is good practice to check for a voltage drop in the fuel pump electrical circuit before replacing the fuel pump. A voltage drop test take all of 5 minutes to do and can save you unnecessary work.

Answer C is wrong, Only Technician B is correct.

Answer D is wrong, Only Technician B is correct.

Question #112
Answer A is wrong. A three-minute charge test is being performed to determine if the battery is still serviceable. A voltage reading over 15.5 volts indicates a defective (sulfated) battery.
Answer B is wrong. Replacing battery electrolyte is not a normal procedure.
Answer C is correct. A voltage reading greater than 15.5 volts during a 3-minute charge test indicates a sulfated battery.
Answer D is wrong. The battery voltage should not exceed 15.5 volt during this test.

Question #113
Answer A is wrong. The voltmeter should not read less than 9.6 volts.
Answer B is wrong. The voltmeter should not read less than 9.6 volts.
Answer C is correct. The voltmeter should not read less than 9.6 volts.
Answer D is wrong. The voltmeter should not read less than 9.6 volts.

Question #114
Answer A is wrong. It is an OBD-II standard.
Answer B is wrong. It is an OBD-II standard.
Answer C is wrong. It is an OBD-II standard.
Answer D is correct. The threshold amount is 1.5 times the limit, not 4.

Question #115
Answer A is wrong. Battery voltage should not drop below 9.6 volts.
Answer B is wrong. Battery voltage should not drop below 9.6 volts.
Answer C is wrong. Neither Technician A nor B are correct.
Answer D is correct. Neither Technician A nor B are correct. Many times the electronic components are misdiagnosed because of low cranking voltage. Cranking voltage must not drop below 9.6 volts while cranking the engine. On some vehicles, if voltage drops below 9.6 volts during cranking there is not enough voltage left to power up crucial components for start-up.

Question #116
Answer A is wrong. Technician B is also correct.
Answer B is wrong. Technician A is also correct.
Answer C is correct. Both Technician A and B are correct. A problem in the alternator such as an open or shorted diode or stator winding will reduce alternator output. High resistance in the alternator field circuit will reduce the magnetic field strength of the rotor and reduce alternator output. Worn brushes or corroded connections would cause high resistance in the field circuit.
Answer D is wrong. Both Technician A and B are correct.

Answers to the Test Questions for the Additional Test Questions Section 6

1. A
2. A
3. C
4. D
5. C
6. A
7. B
8. A
9. A
10. D
11. A
12. C
13. B
14. A
15. A
16. D
17. D
18. D
19. C
20. B
21. B
22. C
23. B
24. C
25. B
26. C
27. C
28. C
29. B
30. A
31. A
32. B
33. A
34. D
35. C
36. D
37. B
38. B
39. C
40. C
41. C
42. D
43. B
44. C
45. D
46. C
47. C
48. B
49. B
50. C
51. C
52. D
53. C
54. A
55. B
56. A
57. A
58. A
59. C
60. A
61. B
62. A
63. C
64. B
65. C
66. B
67. C
68. C
69. A
70. D
71. D
72. B
73. C
74. A
75. B
76. C
77. C
78. A
79. B
80. A
81. B
82. C
83. B
84. A
85. A
86. C
87. C
88. B
89. D
90. A
91. B
92. D
93. C
94. B
95. B
96. C
97. B
98. B
99. D
100. C
101. C
102. D
103. C
104. B
105. A

Explanations to the Answers for the Additional Test Questions Section 6

Question #1

Answer A is correct. Only Technician A is correct. A digital voltmeter (high impedance) can be used to monitor O_2 sensor voltage.

Answer B is wrong. An O2 sensor cannot be checked with diode tester or an ohmmeter.

Answer C is wrong. Only Technician A is correct.

Answer D is wrong. Only Technician A is correct.

Question #2

Answer A is correct. Sulfur odor is usually from a rich condition.

Answer B is wrong. Lack of power is a symptom of low fuel pressure.

Answer C is wrong. Engine surging is a symptom of low fuel pressure.

Answer D is wrong. Limited top speed is a symptom of low fuel pressure.

Question #3

Answer A is wrong. The timing pointer and the timing indicator must be aligned.

Answer B is wrong. An indicator mark aids in the realignment of the distributor.

Answer C is correct. Follow procedures for the specific engine.

Answer D is wrong. When timing a distributor, point the rotor at the distributor cap terminal for the specified cylinder (usually #1).

Question #4

Answer A is wrong. This is a normal condition.

Answer B is wrong. The 02 sensor needs to be kept hot to function properly.

Answer C is wrong. Neither Technician A nor B are correct.

Answer D is correct. Neither Technician A nor B are correct. The secondary air-injection system's function is to supply additional air to the oxidation catalyst to help in the converting of HC and CO to CO2 and water. In the event that the secondary air system becomes inoperative, there will be an increase in HC and CO.

Question #5

Answer A is wrong. Technician B is also correct.

Answer B is wrong. Technician A is also correct.

Answer C is correct. Both Technician A and B are correct. Either a failed crankshaft position sensor or an open ground connection to the DIS assembly would prevent proper primary circuit operation and cause a no-start condition.

Answer D is wrong. Both Technician A and B are correct.

Question #6

Answer A is correct. This signal is for computed spark timing.

Answer B is wrong. The PCM uses this signal.

Answer C is wrong. The PCM uses this signal.

Answer D is wrong. The PCM uses this signal.

Question #7

Answer A is wrong. This is the proper way to adjust mechanical lifters.

Answer B is correct. This statement is not correct about adjusting valves.

Answer C is wrong. The feeler gauge is the proper way to measure clearance.

Answer D is wrong. The piston should be at TDC so that both valves are not under load.

Question #8
Answer A is correct. Worn piston wrist pins create a double knock because the piston changes direction at TDC and BDC every crankshaft revolution.
Answer B is wrong. Excessive timing chain deflection would not create a double knock.
Answer C is wrong. A worn main bearing would not create a double knock.
Answer D is wrong. Excessive main bearing thrust clearance would not create a double knock.

Question #9
Answer A is correct. Only Technician A is correct. OBD-II compliant systems must monitor secondary air injection if present.
Answer B is wrong. Some OBDII compliant vehicles use secondary air-injection systems.
Answer C is wrong. Only Technician A is correct.
Answer D is wrong. Only Technician A is correct.

Question #10
Answer A is wrong. Carbon buildup could cause high readings.
Answer B is wrong. Consistent low readings could be caused by a slipped timing chain.
Answer C is wrong. Low readings on adjacent cylinders usually indicate a blown head gasket. A compression test is performed to check a cylinder's ability to seal and produce compression on the compression stroke. When performing a compression test, if a reading is higher than normal suspect carbon buildup in that cylinder, which would increase the compression ratio for that cylinder. If all the cylinders are low suspect a slipped timing belt or worn cylinder rings. Two adjacent cylinders with low compression could be the result of a blown head gasket. Vacuum or a vacuum leak does not affect compression; however, compression affects vacuum.
Answer D is correct. A vacuum leak affects air/fuel mixture, not cylinder sealing.

Question #11
Answer A is correct. Only Technician A is correct. Use only a digital, high impedance voltmeter.
Answer B is wrong. A test light cannot test an O2 sensor, nor should you try.
Answer C is wrong. Only Technician A is correct.
Answer D is wrong. Only Technician A is correct.

Question #12
Answer A is wrong. Technician B is also correct.
Answer B is wrong. Technician A is also correct.
Answer C is correct. Both Technician A and B are correct. If there is a leak in an adjacent cylinder, look at what is "common" to both cylinders.
Answer D is wrong. Both Technician A and B are correct.

Question #13
Answer A is wrong. A cracked flywheel would not be sensitive to engine temperature and would not diminish when shorting a spark plug wire.
Answer B is correct. Only Technician B is correct. A loose connecting-rod bearing will produce a sharp metallic noise that gets louder as the engine warms up and the oil thins out or when cylinder pressure increases from a throttle snap. Shorting out the cylinders spark plug often makes the noise go away or become much quieter.
Answer C is wrong. Only Technician B is correct.
Answer D is wrong. Only Technician B is correct.

Question #14
Answer A is correct. A scope cannot be used for exhaust gas testing.
Answer B is wrong. A MAP sensor can be tested with an oscilloscope.
Answer C is wrong. A TPS sensor can be tested with an oscilloscope.
Answer D is wrong. A crank position sensor can be checked with an oscilloscope.

Question #15
Answer A is correct. Only Technician A is correct. Overfilling the crankcase can allow the spinning crankshaft to whip up the motor oil causing aeration. Air in the hydraulic lifter will cause the lifter to collapse and the valve train will clatter.
Answer B is wrong. Higher than normal oil pressure will not cause hydraulic lifters to make noise.
Answer C is wrong. Only Technician A is correct.
Answer D is wrong. Only Technician A is correct.

Question #16
Answer A is wrong. Distributor advance does not affect spark strength.
Answer B is wrong. Insulation does not affect spark strength.
Answer C is wrong. Low resistance will not cause weak spark.
Answer D is correct. Added resistance will reduce amount of voltage available to the plug.

Question #17
Answer A is wrong. A restricted exhaust system will result in a continuous drop.
Answer B is wrong. A restricted exhaust system will result in a continuous drop.
Answer C is wrong. Neither Technician A nor B are correct. A restricted exhaust system will not cause fluctuations. One way to check for exhaust restrictions is by using a vacuum gauge. Connect the vacuum gauge to a good manifold vacuum source. Start the engine and record vacuum at idle, then increase the rpm to 2,500 and record the vacuum reading. The reading should be about the same as it was at idle. Allow the throttle to shut rapidly and note vacuum. The vacuum reading should increase about 5 in Hg above the reading at idle then quickly return to record reading at idle. If vacuum at 2,500 rpm shows a continuous gradual drop than suspect a restricted exhaust.
Answer D is correct. Neither Technician A nor B are correct.

Question #18
Answer A is wrong. Worn or tapered cylinder walls increase oil burning and smoke.
Answer B is wrong. Worn valve seals are a common cause of blue exhaust smoke.
Answer C is wrong. Stuck piston rings will cause oil consumption and smoke. When analyzing smoke, grayish/blue smoke is a sign of oil being burned in the combustion chamber. White smoke is a sign of coolant being burned in the combustion chamber. Black smoke is a sign of unburned fuel or a rich running engine.
Answer D is correct. Worn valve seats cause compression loss, not oil consumption.

Question #19
Answer A is wrong. Air loss and bubbles in the radiator indicates a blown head gasket or cracked head or block casting.
Answer B is wrong. Air escaping from the crankcase could indicate problems with the rings.
Answer C is correct. A reading of 100 percent indicates a major, total cylinder leak, either due to an incorrect crank position (valve open) or internal damage. A reading of up to 20 percent is considered normal.
Answer D is wrong. Air escaping from the exhaust indicates a bad exhaust valve. When performing a cylinder leakage test the engine should be at normal operating temperature. The cylinder being tested must be at top dead center of the compression stroke. Less than 10% leakage is good, less than 20% leakage is acceptable, less than 30% leakage is poor, and more than 30% leakage is a problem. Listen for air escaping. Air heard out the tail pipe indicates an exhaust valve leak. Air escaping out the intake manifold indicates an intake valve leak. Air leaking out the radiator indicates a head gasket leak or cracked head or block, and air escaping out the valve cover indicates leaking rings.

Question #20
Answer A is wrong. Throttle bodies do not necessarily have to be removed to be cleaned.
Answer B is correct. Only Technician B is correct. These items need to be verified in addition to performing an idle relearn procedure and clearing any DTC as a result.
Answer C is wrong. Only Technician B is correct.
Answer D is wrong. Only Technician B is correct.

Question #21
Answer A is wrong. A first digit of **P** in an OBD-II trouble code stands for powertrain and is the module that set the code.
Answer B is correct. Only Technician B is correct. An OBD-II code with a 1 as the second digit means the code is a manufacturer specific code, not a generic code.
Answer C is wrong. Only Technician B is correct.
Answer D is wrong. Only Technician B is correct.

Question #22
Answer A is wrong. Technician B is also correct.
Answer B is wrong. Technician A is also correct.
Answer C is correct. Both Technician A and B are correct. Any cylinder that does not produce an rpm drop relative to the other cylinders during a power balance test indicates the cylinder is not producing equal power and has a problem with the ignition, fuel delivery, or mechanical integrity of the cylinder. An inoperative spark plug could cause no rpm drop during a power balance test. A cylinder balance test is performed to check for weak cylinder performance. Cylinder balance on an engine with a distributor is performed by removing the plug wires from the distributor cap one at a time and noting engine rpm drop for each cylinder. On an engine equipped with DIS the cylinders can be cancelled by using a test light and piece of vacuum hose between the coil pack and plug wire. The test light is used to ground each cylinder out. The rpm drops should be within 50 rpm of each other. In the event there is no rpm drop then that cylinder is inoperative due to mechanical or electrical problems.
Answer D is wrong. Both Technician A and B are correct.

Question #23
Answer A is wrong. A loose manifold would not cause the gauge to fluctuate.
Answer B is correct. Only Technician B is correct. Perform a cylinder leakage test to confirm.
Answer C is wrong. Only Technician B is correct.
Answer D is wrong. Only Technician B is correct.

Question #24
Answer A is wrong. Increased fuel delivery is provided through increased injector pulse width.
Answer B is wrong. Vacuum applied to the pressure regulator has no effect on leak down.
Answer C is correct. A vacuum operated pressure regulator provides a constant pressure drop across the injector tip to compensate for changing intake manifold pressure.
Answer D is wrong. Injector spray pattern is controlled by tip and orifice design.

Question #25
Answer A is wrong. There are several variations of the ground symbol.
Answer B is correct. Only Technician B is correct. A ground symbol is made several different ways.
Answer C is wrong. Only Technician B is correct.
Answer D is wrong. Only Technician B is correct.

Question #26
Answer A is wrong. Technician B is also correct.
Answer B is wrong Technician A is also correct.
Answer C is correct. Both Technician A and B are correct. Oil contamination from failing to follow recommended oil change intervals or a restricted oil feed to the turbocharger bearings would both cause repeat turbocharger failure.
Answer D is wrong. Both Technician A and B are correct.

Question #27

Answer A is wrong. Technician B is also correct.

Answer B is wrong. Technician A is also correct.

Answer C is correct. Both Technician A and B are correct. Any interruption of the primary input signals will not switch the module. The crankshaft sensor is the reference signal to the PCM and ignition module, also known as the triggering device. It is this sensor that causes the power transistor to take the primary circuit to ground to allow the coil to build its magnetic field in an electronic ignition system. A defective connection at the crankshaft sensor would result in a no start condition due to no triggering of the ignition module. A defective ignition module would also cause a no start due to no switching of the primary circuit.

Answer D is wrong. Both Technician A and B are correct.

Question #28

Answer A is wrong. Technician B is also correct.

Answer B is wrong. Technician A is also correct.

Answer C is correct. Both Technician A and B are correct. The valve will open late and close early, reducing flow and overlap.

Answer D is wrong. Both Technician A and B are correct.

Question #29

Answer A is wrong. Leaking injectors would cause a rich O2 sensor signal.

Answer B is correct. Only Technician B is correct. The air is not switching, causing O2 sensor signal voltage to remain low. The system will over-richen to compensate, thus the high CO and lean exhaust DTC. Many O2 sensors are replaced due to misdiagnosing the system. In most cases the O2 sensor is doing its job correctly (measuring oxygen content in the exhaust). In this problem, the vehicle is experiencing poor fuel economy and sluggish performance. Both could be caused by a faulty O2 sensor; however, the O2 sensor tests fine. In the event that the air/switching valve (diverter valve) does not switch (sending the secondary air downstream), the secondary air will remain upstream, sending excessive oxygen past the O2 sensor. This will cause the PCM to think the air-fuel mixture is lean and increase the injector pulse width.

Answer C is wrong. Only Technician B is correct.

Answer D is wrong. Only Technician B is correct.

Question #30

Answer A is correct. A digital storage oscilloscope is needed to properly test for slow or lazy oxygen sensors.

Answer B is wrong. A four-gas analyzer can indicate cylinder misfire.

Answer C is wrong. A four-gas analyzer can test catalytic converters.

Answer D is wrong. A four-gas analyzer can pinpoint a plugged fuel injector.

Question #31

Answer A is correct. Only Technician A is correct. The thermostat opening temperature can be monitored on the scan tool and crosschecked with the thermometer.

Answer B is wrong. Accurate operation of the thermostat cannot be checked visually while installed.

Answer C is wrong. Only Technician A is correct.

Answer D is wrong. Only Technician A is correct.

Question #32

Answer A is wrong. The crankshaft must be in the proper position so that the valve lifter is on the base circle of the camshaft, not the cam lobe.

Answer B is correct. Only Technician B is correct. Setting valve clearance too tight can cause the valve to not seat when the engine is warm and combustion pressure will leak past the valve. Rough idle and valve burning will result.

Answer C is wrong. Only Technician B is correct.

Answer D is wrong. Only Technician B is correct.

Question #33
Answer A is correct. Only Technician A is correct. The fan will need to move more air when hotter.
Answer B is wrong. When the coil is cold, the orifice is closed.
Answer C is wrong. Only Technician A is correct.
Answer D is wrong. Only Technician A is correct.

Question #34
Answer A is wrong. A constant high oxygen sensor signal voltage is the result of a rich air-fuel mixture and not the cause. The computer should lean out the air/fuel mixture in response to the high O_2 signal voltage and this would cause lower CO emissions.
Answer B is wrong. An EGR valve not seating would dilute the incoming mixture at idle and cause a misfire and rough idle. The misfire would cause increased hydrocarbons and *lower* CO because combustion does not take place properly with a misfire.
Answer C is wrong. A bad ignition module would not cause an increase in CO emissions, more likely a misfire and increased hydrocarbons.
Answer D is correct. A stuck open purge solenoid could allow fuel vapors to be drawn into the engine during idle, enriching the air/fuel mixture and increasing CO emissions.

Question #35
Answer A is wrong. Poor fuel mileage is a result of improper timing.
Answer B is wrong. If the belt has slipped enough, the engine won't start.
Answer C is correct. Vacuum reading would tend to be lower than normal.
Answer D is wrong. Improper timing can cause a lack of power.

Question #36
Answer A is wrong. A cylinder misfire would cause high emissions.
Answer B is wrong. A lean condition would increase HC and O2.
Answer C is wrong. A faulty fuel-pressure regulator can cause an over-rich condition, increasing CO and HC.
Answer D is correct. A power relay would not affect emissions.

Question #37
Answer A is wrong. Carburetor cleaner creates a mess and is a fire hazard.
Answer B is correct. Propane is best to use when checking for a vacuum leak.
Answer C is wrong. Water can short out electrical components.
Answer D is wrong. Small vacuum leaks cannot be heard.

Question #38
Answer A is wrong. The thermometer would read much higher than ambient temperature.
Answer B is correct. The thermometer should read within a few degrees of thermostat temperature rating.
Answer C is wrong. The temperature reading should be much higher than one-half the thermostat temperature rating.
Answer D is wrong. The temperature reading at the upper radiator hose should be higher than at the lower radiator hose.

Question #39
Answer A is wrong. Shorted armature windings will cause high current draw.
Answer B is wrong. Over-advanced ignition timing will cause high current draw.
Answer C is correct. Corroded battery cables will increase resistance and lower current draw.
Answer D is wrong. Tight engine bearings will cause high current draw.

Question #40
Answer A is wrong. Technician B is also correct.
Answer B is wrong. Technician A is also correct.
Answer C is correct. Both Technician A and B are correct. Electrical components must be removed before soaking the throttle body in solution.
Answer D is wrong. Both Technician A and B are correct.

Question #41
Answer A is wrong. Technician B is also correct.
Answer B is wrong. Technician A is also correct.
Answer C is correct. Both Technician A and B are correct. The dielectric silicone helps transfer heat created in the module to the heat sink surface. Lack of the compound will allow too much heat buildup and can potentially destroy the component.
Answer D is wrong. Both Technician A and B are correct.

Question #42
Answer A is wrong. Low voltage will not harm the ignition switch or starter.
Answer B is wrong. Power to the ignition switch is fused.
Answer C is wrong. Neither Technician A nor B are correct.
Answer D is correct. Neither Technician A nor B are correct.

Question #43
Answer A is wrong. The pickup coil should always be within specification.
Answer B is correct. An open pickup coil winding would show high or infinity on the ohmmeter.
Answer C is wrong. A high resistance reading would indicate an open circuit.
Answer D is wrong. Erratic readings while moving the pickup coil leads indicate a bad connection. Pickup coils are magnetic pulse-generating trigger devices that produce an A/C voltage. These sensors are used to trigger the ignition module to collapse the magnetic field in the ignition coil. Pickup coil resistance is typically 150–1,500 ohms depending on the manufacture. A reading of OL means the pickup coil is open.

Question #44
Answer A is wrong. Technician B is also correct.
Answer B is wrong. Technician A is also correct.
Answer C is correct. Both Technician A and B are correct. The ignition module on many electronic ignition systems controls ignition coil dwell time and current flow.
Answer D is wrong. Both Technician A and B are correct. Dwell is the amount of time in degrees that the primary circuit is taken to ground to build up the magnetic field in the ignition coil. Depending on the manufacture of the vehicle some vehicles' dwell is controlled by the ignition module. By allowing the ignition module to vary the dwell we get more precise coil saturation. Some modules are also equipped with a current limiter to control primary current. If the module does not have a means to limit current there will usually be a primary resister used.

Question #45
Answer A is wrong. Opens in the secondary can cause arc-through damaging windings.
Answer B is wrong. Uncorrected high voltage output requirements can overheat the coil.
Answer C is wrong. If a coil case is cracked, it is defective.
Answer D is correct. An open ignition module primary circuit will not cause an ignition coil failure.

Question #46
Answer A is wrong. Technician B is also correct.
Answer B is wrong Technician A is also correct.
Answer C is correct. Both Technician A and B are correct. A plugged PCV system or excessive crankcase blow-by can both cause oil accumulation in the air filter housing. The function of the PVC system is to vent crankcase pressure from the engine. If the PVC valve were to get restricted or blocked, the first sign of this would be oil leaking from a seal or blowing out the dip stick due to the excessive crankcase pressure. As an engine wears it develops more crankcase pressure, or blowby. On an excessively worn engine the PVC system can't handle all the blow-by so an accumulation of oil ends up in the air cleaner.
Answer D is wrong. Both Technician A and B are correct.

Question #47
Answer A is wrong. Technician B is also correct.
Answer B is wrong. Technician A is also correct.
Answer C is correct. Both Technician A and B are correct. While many ignition systems with computer timing control have the ignition by-pass mode, some late model ignition systems have full PCM control of spark timing.
Answer D is wrong. Both Technician A and B are correct.

Question #48
Answer A is wrong. Changing the cam sensor position will misrepresent cam position.
Answer B is correct. Only Technician B is correct. Some sensors allow for adjustment.
Answer C is wrong. Only Technician B is correct.
Answer D is wrong. Only Technician B is correct.

Question #49
Answer A is wrong. Leaking fuel injectors could cause the symptoms listed.
Answer B is correct. Unmetered air leaking into the engine would cause a lean air/fuel mixture and create high instead of low long-term fuel trim corrections.
Answer C is wrong. A leaking fuel-pressure regulator diaphragm could cause the symptoms listed.
Answer D is wrong. A sticking fuel-pressure regulator could cause the symptoms listed.

Question #50
Answer A is wrong. A bad PCM would affect both injectors.
Answer B is wrong. An open connection would result in the injector not working at all.
Answer C is correct. A shorter voltage spike indicates a shorted injector winding.
Answer D is wrong. A low charging system voltage would affect both injectors. The labscope is used to measure the spike that is left behind when the PCM turns the injector off. Because most injectors are of the electronic type, which means that they are simply a solenoid, they will leave a voltage spike when turned off. We as technicians use this spike to determine the condition of coil of wiring in the injector. A shorted injector winding will cause a low injector spike on a labscope.

Question #51
Answer A is wrong. Technician B is also correct.
Answer B is wrong. Technician A is also correct.
Answer C is correct. Both Technician A and B are correct. Either condition results in added circuit load, causing excessive voltage drop.
Answer D is wrong. Both Technician A and B are correct.

Question #52
Answer A is wrong. This would result in a no-start.
Answer B is wrong. Checking base timing at 2500 rpm would cause the timing to advance.
Answer C is wrong. Neither Technician A nor B are correct.
Answer D is correct. Neither Technician A nor B are correct.

Question #53
Answer A is wrong. Technician B is also correct.
Answer B is wrong. Technician A is also correct.
Answer C is correct. Both Technician A and B are correct. The first step in diagnosing a no-start condition in which the ignition system is suspected is to test for available spark at a spark plug wire. A test light connected to the negative side of an ignition coil that flickers or blinks indicates the primary side ignition components are functioning and the secondary ignition components should be tested next. Both technician statements are valid.
Answer D is wrong. Both Technician A and B are correct.

Question #54
Answer A is correct. It is best to check the fuel pressure and volume first.
Answer B is wrong. There are no fuel pump diagnostic codes.
Answer C is wrong. It is easier to check the fuel pressure and volume first.
Answer D is wrong. Inspecting the fuel lines should only be done if they are suspect. Anytime a faulty fuel pump is suspected the technician should begin by checking the fuel pump's pressure and volume. It's possible for a fuel pump to have good pressure and be low enough in volume to cause a drivability problem.

Question #55
Answer A is wrong. Voltage drop testing is performed to locate high resistance connections. Cleaning connections prior to testing would defeat the purpose of performing the test and prevent the technician from determining if a connection problem was the fault.
Answer B is correct. Only Technician B is correct. Loose or corroded terminals will add unwanted circuit resistance.
Answer C is wrong. Only Technician B is correct.
Answer D is wrong. Only Technician B is correct.

Question #56
Answer A is correct. Only Technician A is correct. The injector off signal voltage is at zero voltage level indicating the coil is grounded and system voltage is applied to turn the injector on. The waveform is inverted by the scope.
Answer B is wrong. A ground-controlled injector would have the injector off signal voltage at 12 to 14 volts, not zero.
Answer C is wrong. Only Technician A is correct.
Answer D is wrong. Only Technician A is correct.

Question #57
Answer A is correct. Only Technician A is correct. The use of a lab scope or DMM can be used to monitor O_2 switching. An oscilloscope is used to measure voltage and time of components such as sensors and output devices such as injectors. An oscilloscope is an excellent way to watch O_2 sensor switching. An oscilloscope can identify O_2 sensor problems such as a lazy O_2 sensor, or a lean or rich biased O_2 sensor, which can be hard to see with just a DVOM. An oscilloscope can also be used to check injector patterns. Using an oscilloscope to check voltage spikes in an injector pattern can identify bad injectors that would otherwise escape detection. The use of an oscilloscope is a way to see a glitch in a component that conventional methods would not reveal.
Answer B is wrong. There are other ways to test fuel injectors, including resistance and injector waveform analysis.
Answer C is wrong. Only Technician A is correct.
Answer D is wrong. Only Technician A is correct.

Question #58
Answer A is correct. Only Technician A is correct. A defective MAP sensor may cause either a rich or lean air-fuel mixture when it fails.
Answer B is wrong. A MAP sensor should be checked by applying vacuum to the sensor with a vacuum pump and graphing the signal output with the key on, engine off. This way sensor output could be checked against the manufacturer's specifications and any engine vacuum problems are eliminated.
Answer C is wrong. Only Technician A is correct.
Answer D is wrong. Only Technician A is correct.

Question #59
Answer A is wrong. Technician B is also correct.
Answer B is wrong. Technician A is also correct.
Answer C is correct. Both Technician A and B are correct. The coolant temperature is used to determine when to change to closed loop during warm-up, as richer mixtures are required.
Answer D is wrong. Both Technician A and B are correct.

Question #60
Answer A is correct. Only Technician A is correct. A damaged air filter will allow dirt to pass through the intake system and cause increased cylinder wall wear.
Answer B is wrong. A restricted air filter will reduce engine breathing and volumetric efficiency. Reduced engine power will decrease fuel economy because the driver will demand more throttle application to maintain acceleration and cruise speeds.
Answer C is wrong. Only Technician A is correct.
Answer D is wrong. Only Technician A is correct.

Question #61
Answer A is wrong. Infinity indicates an open circuit. Any resistance in the circuit will display a reading.
Answer B is correct. Only Technician B is correct. Infinity indicates an open circuit.
Answer C is wrong. Only Technician B is correct.
Answer D is wrong. Only Technician B is correct.

Question #62
Answer A is correct. Only Technician A is correct. Installing a rebuilt alternator in a vehicle with a weak battery causes the alternator to produce high current output trying to recharge the battery. The alternator may overheat and fail.
Answer B is wrong. A battery open circuit voltage (OCV) test only indicates a battery state of charge, not current capacity. A battery can be at 100 percent state of charge but fail a load test and not be fit for service.
Answer C is wrong. Only Technician A is correct.
Answer D is wrong. Only Technician A is correct.

Question #63
Answer A is wrong. Technician B is also correct.
Answer B is wrong. Technician A is also correct.
Answer C is correct. Both Technician A and B are correct. Some vehicles use fuel pump relays and some vehicles don't energize the fuel pump relay until the PCM sees a pulse from the crank sensor.
Answer D is wrong. Both Technician A and B are correct.

Question #64
Answer A is wrong. A dirty oil filter will not cause excessive oil consumption or blue smoke.
Answer B is correct. Worn compressor shaft bearings will cause compressor shaft seal failure and oil will leak into the intake system.
Answer C is wrong. A dirty air filter will not result in oil consumption.
Answer D is wrong. A broken exhaust pipe will not cause oil consumption.

Question #65
Answer A is wrong. Technician B is also correct.
Answer B is wrong. Technician A is also correct.
Answer C is correct. Both Technician A and B are correct. The gas cap is an integral part of the design of the fuel and emission control system, and calibrated for the application. Internal fuel tank pressures can be affected by the use of an improper cap, or a cap that is faulty.
Answer D is wrong. Both Technician A and B are correct.

Question #66
Answer A is wrong. The choke would not be applied when the engine is warm.
Answer B is correct. A worn accelerator pump or misadjusted accelerator pump linkage can cause a hesitation.
Answer C is wrong. A fuel filter that is faulty would cause problems other than a tip-in hesitation.
Answer D is wrong. A fuel return line would cause problems all the time.

Question #67
Answer A is wrong. Technician B is also correct.
Answer B is wrong. Technician A is also correct.
Answer C is correct. Both Technician A and B are correct. If the throttle body is dirty or varnished, it will disturb the airflow into the engine. This can result in an improper throttle plate adjustment. If the throttle plates are adjusted, the angle has changed and the throttle position sensor needs to be adjusted as well.
Answer D is wrong. Both Technician A and B are correct.

Question #68
Answer A is wrong. Answer B is also correct.
Answer B is wrong. Answer A is also correct.
Answer C is correct. Both Technician A and B are correct. Either condition results in a rich mixture.
Answer D is wrong. Both Technician A and B are correct.

Question #69
Answer A is correct. Only Technician A is correct. A plugged or restricted PCV valve or hose will cause crankcase pressure to rise and force oil vapors into the air filter housing.
Answer B is wrong. Only Technician A is correct. Excessive crankcase blow-by is caused by internal engine wear.
Answer C is wrong. Only Technician A is correct.
Answer D is wrong. Only Technician A is correct.

Question #70
Answer A is wrong. Ignition dwell can increase to fully saturate the coil when system voltage is low.
Answer B is wrong. Injector on-time can increase to maintain the proper air/fuel mixture due to slower injector opening when system voltage drops.
Answer C is wrong. The idle speed can increase to help drive the alternator faster at idle speed.
Answer D is correct. Low-system voltage cannot reduce steering effort.

Question #71
Answer A is wrong. Components should not be replaced based solely on a DTC. The trouble code chart will help eliminate other possible causes such as a broken wire or bad connection.
Answer B is wrong. All steps should be followed when using a diagnostic code chart to prevent misdiagnosis. If performing a diagnosis without using a flow chart, the technician must consult manufacturer's specifications to accurately determine the cause of a trouble code.
Answer C is wrong. Neither Technician A nor B are correct. A component should never be replaced solely on a DTC. A diagnostic trouble code only gets the technician on the correct circuit or system. Once the correct system is identified, then a diagnostic flow chart should be used to test the circuit and components properly. A flow chart gives the correct order to test components in order to disclose the correct failure in the system, which may be a component or a wiring problem. When using a flow chart do not skip any steps or a misdiagnosis may occur.
Answer D is correct. Neither Technician A nor B are correct.

Question #72
Answer A is wrong. Restricted passages are always a problem in the exhaust.
Answer B is correct. Only Technician B is correct. EGR passages should always be checked and cleaned when replacing EGR components.
Answer C is wrong. Only Technician B is correct.
Answer D is wrong. Only Technician B is correct.

Question #73
Answer A is wrong. Digital voltmeters can be used when working with PCMs.
Answer B is wrong. Most PCM inputs are low in voltage.
Answer C is correct. Most analog meters are of low impedance design, causing potential damage or incorrect reading in computer circuits.
Answer D is wrong. The oxygen sensor produces voltage from 500 millivolts to 1 volt.

Question #74
Answer A is correct. Only Technician A is correct. 12 volts will be applied through the solenoid and be present at the ECM connector with the solenoid de-energized.
Answer B is wrong. The voltage will be used by the solenoid when energized and be less than .8 volts when measured at the ECM connector.
Answer C is wrong. Only Technician A is correct.
Answer D is wrong. Only Technician A is correct.

Question #75
Answer A is wrong. A saturated charcoal canister or purging the canister during idle can increase tailpipe emissions.
Answer B is correct. A malfunction would usually result in increased emissions.
Answer C is wrong. EVAP system problems may set diagnostic trouble codes and illuminate the MIL.
Answer D is wrong. Fuel odor is a common indication of evaporative system leaks.

Question #76
Answer A is wrong. Technician B is also correct.
Answer B is wrong. Technician A is also correct.
Answer C is correct. Both Technician A and B are correct. Most vehicles will display DTC via a scan tool. Some will display codes by way of pulsing an analog voltmeter connected into the vehicle's diagnostic connector.
Answer D is wrong. Both Technician A and B are correct.

Question #77
Answer A is wrong. Technician B is also correct.
Answer B is wrong. Technician A is also correct.
Answer C is correct. Both Technician A and B are correct. Corrosion or added resistance can occur anywhere in the circuit. A thorough visual inspection and voltage drop tests will help locate problem areas.
Answer D is wrong. Both Technician A and B are correct.

Question #78
Answer A is correct. Only Technician A is correct. A leaking reed valve or aspirator will allow exhaust pressure to back up into the fresh air intake hose. A pulsed secondary air-injection system does not use an air pump. This system uses aspirator valves that operate on the negative pulses of the exhaust. The aspirator valve allows the negative pulses of the exhaust to pull the check off its seat, pulling in fresh air. On the positive pulses of the exhaust the aspirator valve is closed, not allowing any exhaust to escape out the check valve. If a pulsed secondary air-injection check valve gets a leak, exhaust pressure will be felt at the bad aspirator valve.
Answer B is wrong. The oxygen reading on a four-gas analyzer should be used to confirm air-injection operation.
Answer C in wrong. Only Technician A is correct.
Answer D in wrong. Only Technician A is correct.

Question #79
Answer A is wrong. Starter draw is checked in amps.
Answer B is correct. Any "live" circuit will cause amperage flow.
Answer C is wrong. Voltage is measured in volts.
Answer D is wrong. Voltage drops are measured in volts, not milliamps.

Question #80
Answer A is correct. The diverter valve prohibits AIR pump air from entering the exhaust on deceleration. This prevents the continuation of combustion in the exhaust (backfire).
Answer B is wrong. AIR systems do not affect fuel mixture.
Answer C is wrong. Air is not diverted into the passenger compartment.
Answer D is wrong. The AIR system has nothing to do with the A/C system.

Question #81
Answer A is wrong. A faulty mass airflow sensor would Most-Likely cause problems under both cruise and WOT acceleration conditions.
Answer B is correct. Only Technician B is correct. The system may not recognize driver demand or wide-open throttle command if the WOT switch contacts are faulty. This may or may not set a diagnostic trouble code (DTC).
Answer C is wrong. Only Technician B is correct.
Answer D is wrong. Only Technician B is correct.

Question #82
Answer A is wrong. Technician B is also correct.
Answer B is wrong. Technician A is also correct.
Answer C is correct. Both Technician A and B are correct. OBD-II vehicles require a scan tool to retrieve trouble codes; flash codes are no longer supported. OBD-II regulations require the use of standardized diagnostic trouble code formats. On Board Diagnostics generation II systems were mandated by the federal government on all vehicles sold in the United States in 1996 and beyond. Some manufactures were producing OBD II vehicles as early as 1994. With advancement of OBD II vehicles, the ability to retrieve flash codes is no longer supported. A scan tool is required to retrieve codes on an ODB II vehicle. However OBD II regulations require the use of standardized diagnostic trouble codes.
Answer D is wrong. Both Technician A and B are correct.

Question #83
Answer A is wrong. Analog voltmeters may have low input impedance that can cause inaccurate readings.
Answer B is correct. Only Technician B is correct. The digital volt/ohmmeter is the standard for all automotive computer circuit testing.
Answer C is wrong. Only Technician B is correct.
Answer D is wrong. Only Technician B is correct.

Question #84
Answer A is correct. An injector failure will not set a fault code for the EVAP system.
Answer B is wrong. A PCM problem can set a fault code for the EVAP system.
Answer C is wrong. A lack of vacuum to the purge solenoid could set an EVAP code.
Answer D is wrong. A vent or purge solenoid failure in the EVAP system will set a code.

Question #85
Answer A is correct. A defective coil could cause a no-start.
Answer B is wrong. A rotor is located after a coil wire in the circuit.
Answer C is wrong. The primary ignition circuit checks as good.
Answer D is wrong. A diode problem would not cause this failure.

Question #86
Answer A is wrong. Technician B is also correct.
Answer B is wrong. Technician A is also correct.
Answer C is correct. Both Technician A and B are correct. Key off current draw is called a parasitic load and should be less than .05 amps. The alternator supplies the vehicle electrical current when the engine is running under all normal driving conditions. During engine idle with high electrical load the current requirements may exceed alternator output until engine speed increases. The battery supplies the additional current needed.
Answer D is wrong. Both Technician A and B are correct.

Question #87
Answer A is wrong. Technician B is also correct.
Answer B is wrong. Technician A is also correct.
Answer C is correct. Both Technician A and B are correct. Some vehicles will use one sensor input to both engine and body controllers. Improper signals can affect multiple systems.
Answer D is wrong. Both Technician A and B are correct.

Question #88
Answer A is wrong. Cracked or disconnected hoses would result in vapor odor.
Answer B is correct. Only Technician B is correct. If the EVAP system purges vapors from the canister at idle, rough idle will result.
Answer C is wrong. Only Technician B is correct.
Answer D is wrong. Only Technician B is correct.

Question #89
Answer A is wrong. An open field circuit will prevent alternator output.
Answer B is wrong. An open or broken fusible link to the alternator will prevent battery charging.
Answer C is wrong. A broken alternator belt will prevent battery charging. The key word is zero battery charging, meaning no alternator output. An open field circuit in the alternator, burned-out fuse link, and broken belt would all result in no alternator output. One open diode will reduce alternator output but not cause a zero output condition.
Answer D is correct. One open diode will reduce alternator output but not cause a zero charge condition.

Question #90
Answer A is correct. Only Technician A is correct. When checking EGR valve operation one of the first checks to make is to install a hand vacuum pump to the EGR valve and apply 18 inches of mercury to the valve. The engine rpm should drop drastically and idle should become erratic. If so, the EGR valve is functioning properly. Next, the EGR control system should be checked for proper operation. If the EGR valve needs replacement always check all EGR passages for restrictions and blockages.
Answer B is wrong. You should check the EGR passages and control systems, as well as DTCs.
Answer C is wrong. Only Technician A is correct.
Answer D is wrong. Only Technician A is correct.

Question #91
Answer A is wrong. A 2.5 volt drop in a circuit is excessive and a problem exists. The voltmeter leads must be placed across each section of the circuit and the test repeated until the high resistance is found.
Answer B is correct. Only Technician B is correct. This test is an accumulative total of all voltage drops in the circuit. There are several connections and components involved. Each will need to be tested individually to pinpoint the fault.
Answer C is wrong. Only Technician B is correct.
Answer D is wrong. Only Technician B is correct.

Question #92
Answer A is wrong. This checks battery voltage.
Answer B is wrong. This would check resistance.
Answer C is wrong. Neither Technician A nor B are correct.
Answer D is correct. Neither Technician A nor B are correct. To properly check for a draw, an ammeter needs to be used.

Question #93
Answer A is wrong. Technician B is also correct.
Answer B is wrong. Technician A is also correct.
Answer C is correct. Both Technician A and B are correct. If the EGR passages to only one or two cylinders are plugged in the intake manifold, the other cylinders will receive excessive exhaust gas flow when the EGR valve opens and have a density misfire problem. Plugged EGR passages will also raise NOx levels and can result in an IM240 emission test failure.
Answer D is wrong. Both Technician A and B are correct.

Question #94
Answer A is wrong. A specific gravity reading of 1.15 does not indicate a fully charged state.
Answer B is correct. Only Technician B is correct. If the battery gets too hot while recharging, it can be damaged. Refer to specifications for proper charge rate based on battery capacity, state of charge, temperature, and type of charger being used.
Answer C is wrong. Only Technician B is correct.
Answer D is wrong. Only Technician B is correct.

Question #95
Answer A is wrong. Only Technician B is correct. Battery testers cannot be used to measure parasitic loads.
Answer B is correct. Only Technician B is correct. Connect the ammeter in series, so any current draw passes through the meter, to obtain a reading.
Answer C is wrong. Only Technician B is correct.
Answer D is wrong. Only Technician B is correct.

Question #96
Answer A is wrong. If a circuit has excessive resistance, there would not be high current draw.
Answer B is wrong. A defective battery would not cause high current draw.
Answer C is correct. A defective starter would cause high current draw and low cranking speed.
Answer D is wrong. An ignition switch has no relationship to high current draw.

Question #97
Answer A is wrong. The relationship of voltage, amperage and resistance states that excessive amperage and resistance cannot be present at the same time. High resistance in the battery cables will not cause high current draw from the starter motor.
Answer B is correct. Only Technician B is correct. A dragging armature will cause increased current flow in a starter motor and lower cranking speeds.
Answer C is wrong. Only Technician B is correct.
Answer D is wrong. Only Technician B is correct.

Question #98
Answer A is wrong. The windings should always have resistance.
Answer B is correct. The regulator should always have resistance.
Answer C is wrong. Faulty windings would not indicate the condition of the EGR valve.
Answer D is wrong. Faulty windings would not indicate the condition of the MAP sensor.

Question #99
Answer A is wrong. The voltage drop on the positive cable should not exceed 0.5 volts.
Answer B is wrong. The voltage drop on the positive cable should not exceed 0.5 volts.
Answer C is wrong. Neither Technician A nor B are correct.
Answer D is correct. Neither Technician A nor B are correct.

Question #100
Answer A is wrong. Technician B is also correct.
Answer B is wrong. Technician A is also correct.
Answer C is correct. Both Technician A and B are correct. The lower the vacuum applied to the fuel-pressure regulator, the higher the fuel pressure will be. Disconnecting the vacuum line typically raises fuel pressure about 10 psi. If system rest pressure holds after the fuel return line is pinched, it indicates a stuck open or leaking fuel-pressure regulator.
Answer D is wrong. Technician A and B are correct.

Question #101
Answer A is wrong. Technician B is also correct.
Answer B is wrong. Technician A is also correct.
Answer C is correct. Both Technician A and B are correct. This excites the field, causing the alternator to produce near full output. This is basically used as a "process of elimination" when diagnosing charging systems, as it separates the alternator and regulator circuits.
Answer D is wrong. Both Technician A and B are correct.

Question #102
Answer A is wrong. Air pump output is directed to the exhaust manifold during warm-up and switches to the catalytic converter during closed loop operation. Air directed to the exhaust manifold during closed loop operation would create a continuous lean oxygen sensor output.
Answer B is wrong. Air injection causes a large increase in tailpipe oxygen readings when air is directed to the exhaust manifold or catalytic converter.
Answer C is wrong. Neither Technician A nor B are correct.
Answer D is correct. Neither Technician A nor B are correct.

Question #103
Answer A is wrong. Technician B is also correct.
Answer B is wrong. Technician A is also correct.
Answer C is correct. Both Technician A and B are correct. Air-injection diverter valves prevent backfiring during decel by redirecting air pump output to atmosphere. Air-injection check valves allow airflow in only one direction preventing exhaust from backing up into the air-injection system.
Answer D is wrong. Both Technician A and B are correct.

Question #104
Answer A is wrong. A squeal is an indication of a loose belt rather than a growl noise that bearings make.
Answer B is correct. Only Technician B is correct. A loose alternator belt can cause a squeal. A loose belt will almost always emit a noise at start-up. Sometimes a loose belt will emit noise when accelerating from a stop. Humidity will also cause a belt to be noisy. Some serpentine belts will develop a chirp that can't be eliminated without belt replacement. An alternator bearing will emit a growling noise not a squeal.
Answer C is wrong. Only Technician B is correct.
Answer D is wrong. Only Technician B is correct.

Question #105
Answer A is correct. System pinpoint testing is done *after* the complaint is verified.
Answer B is wrong. Always identify the conditions about the complaint.
Answer C is wrong. Road testing a vehicle can verify a complaint.
Answer D is wrong. The repair order contains information that needs to be reviewed.

Glossary

Accelerator A control, usually foot operated, linked to the throttle valve of the carburetor.

Accelerator pump A pump in the carburetor that generates additional fuel to cover for transitions that occur when the throttle position is changed.

Accessory drive As in the belt driven accessories under the hood-fan, alternator, A/C, power steering, air-injection pump.

Air/fuel mixture The proportion of air and fuel supplied to the engine.

Analyzer Any device, such as an oscilloscope, having readout provisions used to troubleshoot a function or event as an aid in making proper repairs.

Automatic choke A system that positions the choke automatically.

Back pressure The excessive pressure buildup in an engine crankcase; the resistance of an exhaust system.

Battery A device used to store electrical energy in chemical form.

Battery cable Heavy wires connected to the battery for positive (hot) and negative (ground) leads.

Battery charger A device used to charge and recharge a battery.

Bearing A device having an inner and outer race with one or more rows of steel balls.

Catalytic converter An exhaust system component to reduce oxides of nitrogen (NO_x), hydrocarbon (HC), and carbon monoxide (CO).

Check valve A device that allows the flow of liquid or vapor in one direction and blocks it in the other direction.

Coil That part of the ignition system that provides high voltage for the spark plugs.

Cold-cranking amperage The number of amperes that a fully charged battery will provide for 30 seconds without the terminal voltage dropping below 7.2 volts.

Combustion chamber The area above a piston, at top dead center, where combustion takes place.

Compression The process of squeezing a vapor into a smaller space.

Compression test A measurement of the pressure a cylinder is able to generate during a controlled cranking period.

Computer A system capable of following instructions and to alter data in a desirable way in order to perform operations without human intervention.

Cooling system The radiator, hoses, heater core, and cooling jackets used to carry away engine heat and dissipate it in the surrounding air.

Cruise control A system of automatically maintaining preset vehicle speed over varying terrain.

Customer complaint The description of a problem provided by the customer, usually the driver of the vehicle.

Cylinder balance A dynamic test that shorts out the engine cylinders one at a time and compares the power loss in each to pinpoint weak cylinders.

Cylinder head That part that covers the cylinders and pistons.

Cylinder leakage test A test to determine how well a cylinder seals when the piston is at top dead center and the valves are closed.

Deck The flat mating surfaces of an engine block and head.

Dedicated ground There are many ground connections on an automobile, some are dedicated to a particular component or circuit.

Diaphragm A flexible rubber-like membrane.

Digital ohmmeter A device that sends a small amount of current into an isolated circuit and indicates the amount of resistance in a numerical readout.

Digital voltmeter A device that reads the difference in voltage pressure at two points of an electrical circuit in a numerical readout.

Driveability A term used for any problem or compliant the driver might encounter in the engine control system or transmission control system.

Driveability problem A problem or compliant encountered in the engine control system or transmission control system.

Drive belt The belt or belts used to drive the engine-mounted accessories off the crankshaft.

Dwell The degree of distributor shaft rotation while the points are closed.

EGR An abbreviation for exhaust gas recirculation valve. A valve that meters a small amount of exhaust gas into the intake manifold during light cruise conditions to lower combustion chamber temperatures and reduce the formation of nitrogen oxides.

EGR valve A valve that meters a small amount of exhaust gas into the intake manifold during light cruise conditions to lower combustion chamber temperatures and reduce the formation of nitrogen oxides (NO_x).

Electronic control A control device that is electrically or electronically actuated.

Electronics Pertaining to that branch of science dealing with the motion, emission, and behavior of currents of free electrons.

Emission A product, harmful or not, emitted to the atmosphere. Emissions are generally regarded as harmful.

Emission control Emission tests are tests done on the vehicle to determine how it compares to federal government standards for tail pipe emissions, crankcase emissions and evaporative emissions.

Emission control Components that are directly or indirectly responsible for reducing harmful emissions.

Emission test The use of calibrated equipment to determine the amount of emissions that are being released to the atmosphere.

Engine A prime mover. A device for converting chemical energy (fuel) to useable mechanical energy (motion).

Engine-management system An electronic system that monitors, regulates, and adjusts, engine performance and conditions.

Engine manifold vacuum The vacuum signal taken directly off the intake manifold or below the throttle plate.

Engine oil A lubricant formulated and designated for use in an engine.

Exhaust The byproduct of combustion; the pipe from the muffler to the atmosphere.

Exhaust back pressure The pressure that develops in the exhaust system during normal operation. Two pounds is cause for concern.

Exhaust gas recirculation valve A valve that meters a small amount of exhaust gas into the intake manifold during light cruise conditions to lower combustion chamber temperatures and reduce the formation of nitrogen oxides.

Exhaust port An opening that allows the exhaust gases to escape.

F An abbreviation for Fahrenheit, a temperature measurement in the English scale.

Fail-safe A default mode designed into many operating systems that allows limited function of a system when a malfunction occurs. This is to protect the system or to allow the driver to move the vehicle to a safe area.

Fan blade A flat pitched part of a fan that moves air.

Fault code A numeric readout system used as an aid in troubleshooting procedures, information about functions or malfunctions of an electronic system.

Fuel contamination Any impurities in the fuel system.

Fuel injector Electrical or mechanical devices that meters fuel into the engine.

Fuel pressure The pressure of the fuel in an injected or non-injection fuel system.

Fuel-pressure regulator A device that regulates fuel pressure. Fuel injected engines require pressure regulators because some fuel pumps develop over 100 psi (690 kPa).

Fuel pump An electrical or mechanical device that pumps fuel from the tank to the carburetor or injection system.

Fuel volatility A term used to determine how rapidly a fuel evaporates or burns.

Hand choke A choke that is cable controlled manually.

Head gasket A sealing material between the head and block.

Hot lines Special telephone or computer lines for information access for help with problem solving.

Idle speed The speed of an engine at idle with no load.

Ignition system The system that supplies the high voltage required to fire the spark plugs.

Ignition timing The interval, in crankshaft degrees of rotation, before top dead center that a spark plug fires.

Intake manifold That part of the engine that directs the air fuel mixture to the cylinders.

Intake port That part of a cylinder, having a valve, that allows the air fuel mixture to enter the combustion chamber.

Jumped A term often used in reference to timing chains, belts, or gears; meaning valve timing is off, or out of specifications. This happens when parts wear or loosen.

Jump start To aid starting by the use of an external power source, such as a battery or battery charger.

Keep alive memory A program in many computerized devices that retains fault code information and other information necessary for the operation of the system.

Key off battery drain A term used for parasitic drains. Electrical demands on the battery while the ignition key is in the off position.

Knock sensor A sensor that signals the engine control computer when detonation is detected retarding ignition timing.

LFT: Long-term fuel trim "Permanent" addition or subtraction of fuel assignment for fuel injected vehicles.

Lubrication The act of applying lubricant to fittings and other moving parts.

Magnaflux A dry, nondestructive magnetic test to check for cracks or flaws in iron or steel parts.

Main bearing The bearing that supports the crankshaft in the lower end of an engine.

Module A semi-conductor device designed to control various systems like, ignition, engine control, steering, suspension, brakes, transmissions, power windows, power seats, windshield wipers, brakes, traction control, and cruise control.

Oil A lubricant.

Oscilloscope An instrument that produces a visible image of one or more rapidly varying electrical quantities with respect to time and intensity.

Oxygen sensor A device located in the exhaust system close to the engine that reacts to the different amounts of oxygen present in the exhaust gases, and sends signals the engine control computer so it can maintain the proper air/fuel ratio.

PCV An abbreviation for positive crankcase ventilation. A metering device connecting the engine crankcase to engine vacuum, which allows burning of crankcase vapors to reduce harmful engine emissions.

PCV valve A metering device connecting the engine crankcase to engine vacuum, which allows burning of crankcase vapors to reduce harmful engine emissions.

Pings Unscheduled explosions in the combustion chamber. Also referred to as spark knock or detonation.

Positive crankcase ventilation valve A metering device connecting the engine crankcase to engine vacuum, which allows burning of crankcase vapors to reduce harmful engine emissions.

Power balance A dynamic test that shorts out one engine cylinder at a time and compares the power loss to pinpoint weak cylinders.

PSI (psi) Abbreviation for pounds per square inch. Used in the English system for air, vacuum, and fluid pressure measurements.

Rev limiter A device that limits the revolutions (turning speed) of a device or component.

Scanner A device, usually hand held, that accesses the electronic systems on vehicles to obtain fault codes and operating parameters. Can be used to simulate signals and to verify operation of systems on some vehicles.

Scan tool A tester used to recall trouble codes.

Seal A gasket-like material between two or more parts or a ring-like gasket around a shaft to prevent fluid or vapor leakage.

Secondary air insertion system Outside air that is pumped into the exhaust system and the catalytic converter to promote continued burning and chemical reactions that reduce harmful exhaust gas emissions.

Sensor An electrical sending unit device to monitor conditions for use in controlling systems by a computer.

Serpentine belt A wide flat belt with multiple grooves that winds through all of the engine accessory pulleys and drives them from the crankshaft.

Severe service Any vehicle service beyond average conditions, such as a taxi cab.

Servo A device that converts hydraulic pressure to mechanical movement.

SFT: Short-term fuel trim Corrects short term exhaust conditions and load.

Shroud A hood-like device used to direct air flow.

Speed limiter Usually a program in the engine control computer designed to limit the speed of the vehicle because of tire speed rating.

Supercharged A belt driven device that pumps air into the engine induction system at a pressure higher than atmospheric pressure.

Supercharger A belt driven device that pumps air into the engine induction system at a pressure much higher than atmospheric pressure.

Tail pipe The tube-like components that directs exhaust vapors from the outlet of the muffler to the atmosphere.

Technical information Information found in manufacturers manuals, bulletins, reports, text books, and other such sources.

Technical service bulletin Periodic information provided by the manufacturer relative to production changes and service tips.

Temperature The heat content of matter as measured on a thermometer.

Timing The spark delivery in relation to the piston position.

Trouble code Numbers generated by the diagnostic system that refer to certain troubleshooting procedures.

TSB An abbreviation for Technical Service Bulletins. TSBs are made available by product manufacturers to make known product improvements, problems found, fixes, revisions, and safety related concerns.

Turbocharger A device, driven by exhaust gases, that pumps air into the engine induction system at a pressure higher than atmospheric pressure.

Unmetered Matter, such as air, that is entering a controlled area without being measured by a management system.

Unmetered air Air that is entering a controlled area without being measured by the air management system.

Valve timing The opening and closing of valves in relation to crankshaft rotation.

V-belt A rubber-like V-shaped belt used to drive engine mounted accessories off the crankshaft pulley or an intermediate pulley.

Verify the repair To retest a system and/or test drive the vehicle after repairs are made.

Voltage A quantity of electrical force.

Voltage spikes Higher than normal voltage often caused by collapsing magnetic fields.

Wastegate A device on superchargers and turbochargers that limits the amount of pressure increase in the intake to safe design limits.

Water pump A device, usually engine driven, for circulating coolant in the cooling system.

Wrap A term used for the amount of contact a belt has on a pulley.

Notes

Notes

Notes

Notes

Notes

Notes

Notes

Notes

Notes

Notes

Notes

Notes

Notes